AF526782

Manual of Practical Biochemistry

FIFTH EDITION

Manual of Practical Biochemistry

FIFTH EDITION

As per the CBME Curriculum (NMC)

RAFI MD

Dean and Medical Director
Surabhi Institute of Medical Sciences
Siddipet, Telangana

Universities Press

All rights reserved. No part of this book may be modified, reproduced or utilised in any form, or by any means, electronic or mechanical, including photocopying, recording or by any information storage and retrieval system, in any form of binding or cover other than in which it is published, without permission in writing from the publisher.

MANUAL OF PRACTICAL BIOCHEMISTRY, FIFTH EDITION

UNIVERSITIES PRESS (INDIA) PRIVATE LIMITED

Registered Office
3-6-747/1/A & 3-6-754/1, Himayatnagar, Hyderabad 500 029, India
info@universitiespress.com: www.universitiespress.com

Distributed by
Orient Blackswan Private Limited

Registered Office
3-6-752, Himayatnagar, Hyderabad 500 029, India

Other Offices
Bengaluru / Chennai / Guwahati / Hyderabad
Kolkata / Mumbai / New Delhi / Noida / Patna

© Rafi MD 2014, 2020, 2023, 2025
First edition published by Pathfinder Medical Publishers 2013
Second edition published by Universities Press 2014
Reprinted 2015, 2018
Third edition 2020
Fourth edition 2023
Fifth edition 2025

ISBN: 978 81 974249 7 7

502140

Cover and book design
© Universities Press (India) Private Ltd 2014, 2020, 2023, 2025

Typeset in Minion Pro 10.5/12.5 *by*
Bookcraft Publishing Services (I) Pvt Ltd, Chennai

Printed in India by
SDR Printers , Tronica City, Ghaziabad 201 102

Published by
Universities Press (India) Private Limited
3-6-747/1/A & 3-6-754/1, Himayatnagar, Hyderabad 500 029, India

Care has been taken to confirm the accuracy of information presented in this book. The author and the publisher, however, cannot accept any responsibility for errors or omissions or for consequences from application of the information in this book and make no warranty, express or implied, with respect to its contents.

Contents

A Mapping of the Contents to the CBME Practical Biochemistry Curriculum

Competency number	Description	Chapters in the book
BI 11.1	Describe commonly used laboratory apparatus and equipment, good safe laboratory practice and waste disposal.	**1, 5, 25**
BI11.2	Describe the preparation of buffers and estimation of pH.	**24**
BI11.3	Describe the chemical components of normal urine.	**3**
BI11.4	Perform urine analysis to estimate and determine normal and abnormal constituents.	**3, 4**
BI11.5	Describe screening of urine for inborn errors and describe the use of paper chromatography.	**22, 28**
BI11.6	Describe the principles of colorimetry.	**5**
BI11.7	Demonstrate the estimation of serum creatinine and creatinine clearance.	**9**
BI11.8	Demonstrate estimation of serum proteins, albumin and A:G ratio.	**11**
BI11.9	Demonstrate the estimation of serum total cholesterol and HDL cholesterol.	**12**
BI11.10	Demonstrate the estimation of triglycerides.	**12**
BI11.11	Demonstrate estimation of calcium and phosphorous.	**18**
BI11.12	Demonstrate the estimation of serum bilirubin.	**13**
BI11.13	Demonstrate the estimation of SGOT/ SGPT.	**14**
BI11.14	Demonstrate the estimation of alkaline phosphatase.	**15**
BI11.15	Describe and discuss the composition of CSF.	**20**
BI11.16	Observe the use of commonly used equipment/techniques in biochemistry laboratory including: – pH meter – paper chromatography of amino acid – protein electrophoresis – TLC, PAGE – electrolyte analysis by ISE	**5, 19, 21, 22, 26, 27**

Competency number	Description	Chapters in the book
BI11.16	– ABG analyser – ELISA – immunodiffusion – autoanalyser – quality control – DNA isolation from blood/ tissue	
BI11.17	Explain the basis and rationale of biochemical tests done in the following conditions: – diabetes mellitus – dyslipidemia – myocardial infarction – renal failure – gout – proteinuria – nephrotic syndrome – edema – jaundice – liver diseases – pancreatitis – disorders of acid–base balance – thyroid disorders	**2, 5, 6, 9, 16, 17, 27**
BI11.18	Discuss the principles of spectrophotometry.	**5**
BI11.19	Outline the basic principles involved in the functioning of instruments commonly used in a biochemistry laboratory and their applications.	**5**
BI11.20	Identify abnormal constituents in urine, interpret the findings and correlate these with pathological states.	**4**
BI11.21	Demonstrate estimation of glucose, creatinine, urea and total protein in serum.	**6, 7, 8, 9, 11**
BI11.22	Calculate albumin: globulin (A/G) ratio and creatinine clearance.	**9, 11**
BI11.23	Calculate the energy content of different food items, identify food items with high and low glycemic index and explain the importance of these in diet.	**29**
BI11.24	Enumerate advantages and/or disadvantages of use of unsaturated, saturated and trans fats in food.	**30**

Preface to the Fifth Edition

I am pleased to present the fifth edition of *Manual of Practical Biochemistry*. The fourth edition, published two years ago, has received positive responses from various parts of the country. A number of valuable suggestions have also been received, which have contributed greatly to the development of this edition. The revised contents covers all the topics prescribed in the CBME curriculum.

This edition has been reorganised as per the revised syllabus for practical biochemistry prescribed in the CBME curriculum. A few chapters on qualitative analyses have also been deleted.

Chapters on urinary screening for inborn errors of metabolism, energy content of various foods and their glycemic index, and saturated, unsaturated and trans-fatty acids in foods are included. As in the previous editions, commonly-asked viva questions are included at the end of respective chapters to help students in their final practical examination preparations. Case studies are dispersed throughout the chapters in addition to the selection included in Appendix III.

Acknowledgements

I express my sincere thanks to Dr Ayesha Jabeen, MD, Associate Professor, MIMS, Hyderabad, and Dr Praveen Kumar, Surabhi Institute of Medical Sciences (SIMS), Siddipet, for their valuable contributions to this edition. Additionally, I am grateful to Universities Press, particularly to Mr Madhu Reddy, Director, for his encouragement and support, to Dr Sudha Ganesan, who helped with the revision plan, and to Sreelatha Menon for her editorial assistance. My thanks to Dr K Praveen Kumar, CEO of SIMS for his support and to Ms Shravanthi for her technical assistance. To my family, I express my gratitude for encouragement and support, as always.

I sincerely appreciate feedback from readers to improve the book further.

September 2024
Hyderabad

Rafi MD
drrafiauthor@gmail.com

Preface to the First Edition

As the name implies, this book is intended for students pursuing medicine and allied sciences. It has been written strictly adhering to the syllabus and guidelines prescribed by the Medical Council of India (MCI). The conventional teaching of practical biochemistry is more chemistry-oriented than focused on clinical applications. In the context of rapid expansion in the science of clinical biochemistry all over the world, the traditional practical methods are slowly being replaced by new clinical analyses in medical schools.

This book presents the principles of practical and clinical biochemistry in a lucid style and an easily understandable form. Apart from discussing the principles and methodology of estimations of various biochemical parameters, it stresses on their applied aspects. The book is organised into qualitative analyses and quantitative estimations, covering the principles, procedures and observations, followed by clinical interpretation.

Specially prepared questionnaires in the form of MCQs, problem-based exercises, interpretation of laboratory data and case studies are incorporated at the end of the book as appendices. This provides valuable material to the students preparing for various competitive examinations in India and abroad. It has been observed by the author that most of the medical students are not confident of handling problems pertaining to physical chemistry. To compensate for this, a few problem-solving exercises have been added at the end of the book as part of an appendix, discussing the principles and concepts of molarity, normality and preparation of different types of solutions, etc.

Though it has been written primarily keeping in view of the needs of medical students, it is also useful to students of allied sciences including Dental, Pharmacy, Physiotherapy, Nursing, MLT, Biotechnology, Biomedical Engineering, Agriculture, Veterinary, Biology and Microbiology. I thank Mr E. Vijay of Dharani Graphics, Hyderabad, for composing and designing the book. I also express my sincere thanks to Mr. Vijay and Mr. Umakanth of Shiridi Sai Graphics, Hyderabad, for composing a few parts of this book.

I have done my best to make this manual a complete practical guide for all the student needs. I hope it will serve as a valuable companion throughout their career. Opinions and suggestions from both students and the teaching community will be acknowledged and honoured.

April 2013
Hyderabad

Rafi MD
drrafiauthor@gmail.com

To my wife, Tikku,
for her enduring encouragement, support and love

1

BI 11.1

Safety in the Laboratory

Students must be careful while working in the laboratory with chemicals, glassware or samples to be tested, as they are potentially dangerous if wrongly handled.

Precautions

- Be cautious while using chemicals/reagents because if splashed on the skin or eyes they may cause injury. Do not inhale any vapours produced during experiments.
- Do not carry large bottles by the neck. Hold them firmly with both hands.
- Flammable chemicals like glacial acetic acid, methanol, ethanol, acetone and acetic acid should be kept away from flames. Do not heat flammable liquids over a flame; instead, use a water bath or an electric hot plate.
- Always use protective gloves while handling toxic chemicals. Wash your hands after using a toxic chemical. **Mouth pipetting of toxic chemicals is totally forbidden**. All toxic chemicals—for example, potassium cyanide, sodium nitroprusside, formaldehyde and sodium azide—must be under safe custody (locked in a cupboard).
- Never perform mouth-pipetting of corrosive chemicals like concentrated H_2SO_4, HCl, nitric acid, trichloroacetic acid and glacial acetic acid. Always add the corrosive chemical to water very slowly and not vice versa.
- Certain chemicals like benzidine (used to detect blood in urine) and O-toluidine (used to measure glucose concentration in blood) are carcinogenic (cancer causing) and should be handled with utmost care, avoiding skin contact, inhalation or ingestion. Never add water directly to acid because water spurts, generating a lot of heat, causing the acid to splash out. Addition of acid to water should be done in small quantities, accompanied by constant stirring. Highly concentrated acids should be diluted in a chest containing ice cubes.

First Aid in the Laboratory

Acid Burns

Wash immediately with large quantities of water and do the following.

In case acid splashes on the skin

- Wash thoroughly and repeatedly with water.
- Clean the affected area with cotton wool soaked in 5% aqueous Na_2CO_3.

In case acid splashes in the eye

- Wash the eye with water sprayed from a bottle. Rinse thoroughly, especially the corner of the eye near the nose. Otherwise, wash the eye out under running water.

- Apply 4–6 drops of 2% aqueous sodium bicarbonate to the eye. Continue applying it till a doctor attends to it.

In case of accidental swallowing of acids

Acid may be swallowed accidentally while pipetting.

☞ **Contact a physician immediately or shift the person to the Casualty Department of a nearby hospital.**

- Meanwhile, make the person drink a 5% soap solution or eat two egg whites mixed with half a litre of water or milk.
- If nothing is available, the person should drink plenty of water.
- Rinse the lips and tongue thoroughly with water and 2% aqueous sodium bicarbonate.
- Let the person gargle with the soap solution.

Alkali Burns

In case alkali splashes on the skin

- Wash thoroughly and repeatedly with water.
- Clean the affected skin with cotton soaked in 5% acetic acid.

In case alkali splashes in the eye

- Wash the eye with water thoroughly—bathe it under running tap water.
- Wash the eye with saturated boric acid solution.

In case of accidental swallowing of alkali while pipetting

- Make the person drink lemon juice or diluted vinegar (1:4 dilution with water). Make the person gargle with 5% acetic acid.
- In case the alkali burn affects the lips and tongue, wash well first with water and then with 5% acetic acid solution.

Fire Accidents

Fire accidents may occur due to the presence of electrical appliances, spirit lamps, flammable gases stored in cylinders, etc., and hence all laboratories must have a fire extinguisher. Smoking must be prohibited in the laboratory.

First Aid Box

It is mandatory for every laboratory to have a **first aid box**. It should contain the following items:

- Sterilised gauze and cotton
- Sterilised eye pads
- Wash bottle for washing eyes
- Bandage of 2" width, crepe bandage
- Injection adrenaline (1 in 1000) – 2 ampoules
- Injection aminophyline (250 mg in 100 mL) – 2 ampoules
- Dexona – 2 vials
- Disposable syringes and needles of different capacities
- 1% acetic acid (1 bottle)
- Milk of magnesia (1 bottle)
- Adhesive plaster, scissors
- Soap powder solution (5 g per litre of water)
- Antiseptic lotion

2 Reactions of Non-protein Nitrogenous (NPN) Substances

BI 11.3

The nitrogen containing substances in the body, other than proteins and nucleic acids, are known as **non-protein nitrogenous (NPN) substances**. They include urea, uric acid, creatine, creatinine and ammonia. Ammonia is converted to urea in the body before being eliminated. Usually the reactions of urea, uric acid and creatinine are performed as reactions of NPN.

Urea

Brief notes on urea

- Urea is the end product of amino acid metabolism.
- The nitrogen of amino acids (amino group) is converted to ammonia (transamination and deamination).
- Ammonia is potentially toxic at elevated levels and hence is converted to less toxic urea in the liver.
- Urea is synthesised in the liver through a series of reactions known as the 'urea cycle'.
- From the liver, urea is transported to the kidneys and excreted in urine.
- Normal blood urea level is 10–40 mg/dL.

Elevated urea levels in blood (**uremia**) is seen in the following conditions.

1. **Pre-renal** causes → Dehydration (diarrhea and vomiting)
2. **Renal** causes → ARF (acute renal failure) CRF (chronic renal failure)
3. **Post-renal causes** → Renal calculi and enlargement of prostate

Reactions of Urea

1. Sodium hypobromite test

This is a test used to detect urea.

Principle

When urea is treated with sodium hypobromite, it decomposes to liberate nitrogen. Released nitrogen gas produces marked effervescence.

Procedure

1. Take 3 mL of urea solution and add 4–6 drops of sodium hypobromite.

Observation

Brisk effervescence is seen.

Inference

When urea is treated with sodium hypobromite, it decomposes to release nitrogen gas, which produces brisk effervescence.

2. Specific urease test

This is the specific test for urea. *Urease* is an enzyme that occurs in horse gram, jack beans and watermelon seeds.

Principle

When urea is treated with the enzyme urease, it is converted to ammonium carbonate which is alkaline (pH is raised). At this raised pH, phenophthalein indicator shows a pink colour (pink colour in alkaline medium).

Procedure

1. Take two test tubes and label them 'T' (test) and 'C' (control).
2. Take 5 mL of urea solution in both the test tubes. Add 2 mL of urease suspension to both the tubes and heat the control tube.
3. Incubate the tubes at room temperature for 15 minutes.
4. Add 2 drops of phenolphthalein indicator to both.

Observation

Observe the development of pink colour in the 'T' test tube and no reaction in 'C' test tube.

Inference

The enzyme urease acts on urea in the 'T' test tube to liberate ammonia and carbonic acid. Under the pH conditions of the reaction, they are converted to ammonium carbonate which raises the pH to 8.5 since it is alkaline. At this pH, phenolphthalein gives a pink colour. The solution in the 'C' test tube will not give any colour because when heated, the enzyme urease is denatured and inactivated and hence no reaction occurs.

Uric acid

Brief notes on uric acid

1. Uric acid, another NPN, is an end product of purine catabolism.
2. Normal value of uric acid in blood is 2–6 mg/dL.
3. Elevated levels of uric acid (**hyperuricemia**) are associated with **gout** and **renal failure**.

Reactions of Uric Acid

1. Benedict's uric acid test (or) phosphotungstic acid reduction test

Uric acid is a strong reducing agent in alkaline conditions.

Principle

Uric acid, being a reducing agent, reduces phosphotungstate to tungsten blue in an alkaline medium.

Procedure

1. Take 2 mL of uric acid solution, add a few drops of phosphotungstic acid reagent.
2. Then add a few drops of 20% Na_2CO_3.

Observation

Observe the emergence of blue colour.

Inference

Uric acid reduces phosphotungstic acid to tungsten blue in alkaline conditions, giving rise to the blue-coloured solution.

2. Schiff's test

This is the specific reaction for uric acid.

Principle

Uric acid reduces ammonical $AgNO_3$ to metallic silver.

Procedure

1. Take a filter paper and wet it with two drops of ammoniacal $AgNO_3$ solution.
2. Add 1–2 drops of uric acid solution to the same filter paper.

Observation

Observe the development of black colour on the filter paper due to the precipitation of silver.

Inference

Uric acid reduces ammonical $AgNO_3$ to metallic silver, turning it black.

3. Murexide test

This is the specific test for uric acid.

Principle

When uric acid is treated with concentrated HNO_3, it oxidises to give purpuric acid (reddish yellow). This, in turn, combines with ammonia to form purple-red–coloured ammonium purpurate (*murexide*).

Procedure

1. Take 5 mL of uric acid solution in a china dish, add 4–5 drops of concentrated HNO_3.
2. Evaporate over a low flame cautiously. A red residue is obtained.
3. Allow the china dish to cool. Then add 1–2 drops of dilute ammonia.

Observation

Note the formation of a purplish red colour. (Colour disappears on warming).

Inference

When uric acid is treated with concentrated HNO_3, it undergoes oxidation. The ring structure is cleared and the derivatives condense to give purpuric acid. The ammonium salt of purpuric acid (murexide) is purplish red in colour.

Creatinine

Brief notes on creatinine

- Creatinine, another NPN substance, is derived from creatine.
- Creatine is synthesised from the amino acids glycine, arginine and methionine.
- Creatine exists as creatine phosphate in muscle tissue. **Creatine phosphate** acts as the storage form of high-energy phosphate. It replenishes the ATP in muscles. Creatine phosphate transfers phosphate moiety to ADP, converting it to ATP. This is an important reaction during muscle contraction and is known as '**Lohmann's reaction**'.
- The normal concentration of creatinine in blood is 0.6–1.5 mg/dL.
- Creatinine levels are elevated in renal insufficiency and renal failure.
- Creatinine is excreted into urine at the rate of 1–2 g/day.
- Estimation of serum creatinine levels is considered a reliable diagnostic indicator to assess **kidney function**.

Reactions of Creatinine

Jaffe's test

This is the specific test used to detect and estimate creatinine levels.

Principle

Creatinine reacts with picric acid in an alkaline medium to form creatinine picrate (an orange-red–coloured product).

Procedure

1. Take two test tubes and label them 'T' (test) and 'C' (control).
2. In the 'T' test tube, take 5 mL of creatinine solution and 2 mL of saturated picric acid and a few drops of 10% NaOH. Shake well and keep for a few minutes.
3. In the 'C' test tube, take 5 mL of water and 2 mL of picric acid and a few drops of 10% NaOH. Shake well and keep for a few minutes.

Observation

Note the orange-red colour in the 'T' tube and the yellow colour in the 'C' tube.

Inference

Creatinine reacts with alkaline picrate solution to form orange-red creatinine picrate. Creatine can also be identified by this test as it is easily converted to creatinine.

Viva Questions

1. What is meant by the term '*NPN substances*'?
2. What are the tests used to detect urea in a sample? In the hypobromite test, the effervescence is due to which gas?
3. What is the action of the enzyme *urease*? Does the human body synthesise urease?
4. Urea is the end product of which metabolism? What is the normal blood urea level?
5. How is urea removed from the body?

6. Name the test that is used to detect creatinine in the sample.
7. Creatinine is synthesised from which amino acids?
8. What is the difference between creatinine and creatine?
9. What is *Lohman's reaction*? What is the role of creatine phosphate?
10. What is the clinical significance of the enzyme *creatine kinase*?
11. Uric acid is the end product of which metabolism?
12. Name the tests used to detect uric acid in samples.
13. What is the normal serum level of uric acid?
14. Is uric acid a reducing agent?

Identification of Unknown NPN Solution

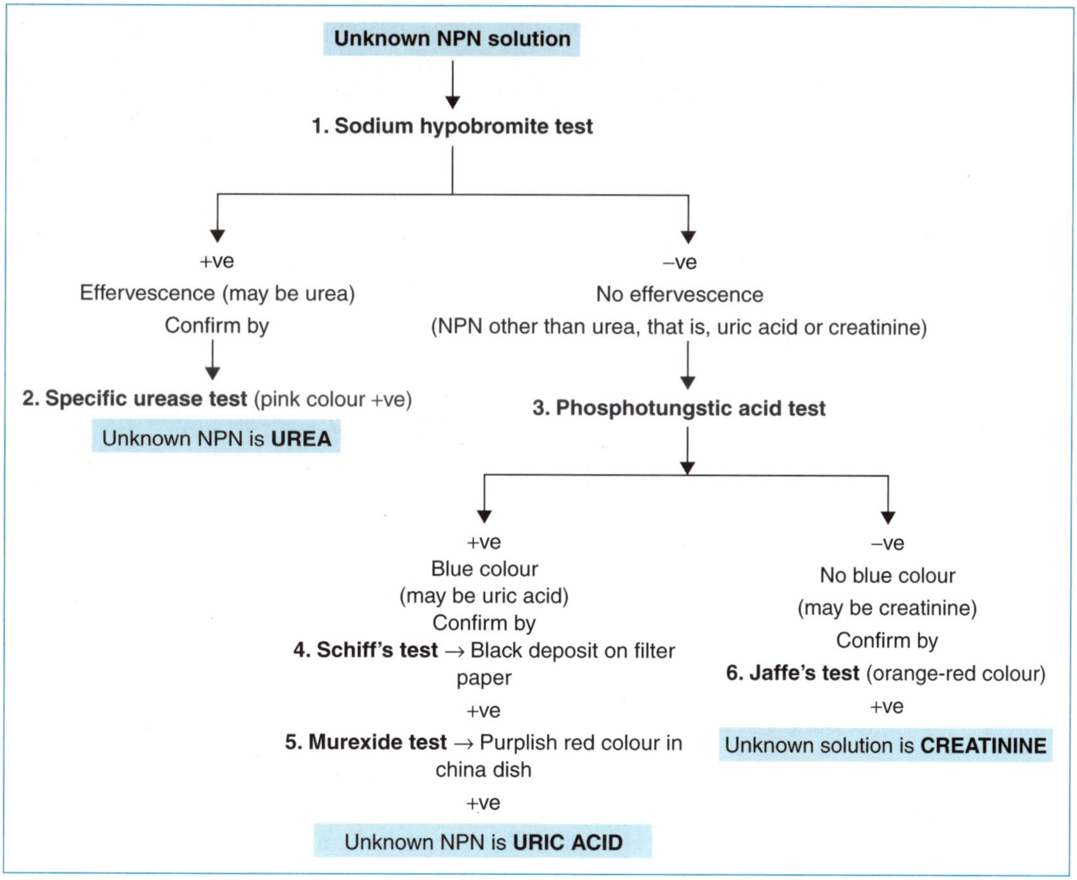

3 Analysis of Urine

BI 11.3; 11.4

Urine analysis (**urinalysis**) is a routine investigation in clinical practice because it gives valuable information which aids the physician in arriving at a proper diagnosis.

How to examine urine?

Urine should be examined preferably soon after being passed. Usually a random sample (collected at any time of the day) is used, except on some special occasions (for example, to detect sugar in urine, a sample collected two hours after meals is preferred). For estimating certain biochemical constituents like urea, uric acid, creatinine and proteins, a 24-hour urine sample is used with a preservative (for example, thymol) to prevent bacterial growth.

Physical Examination

Under this heading, we examine the volume, colour, pH and specific gravity of urine.

Volume

A normal healthy adult excretes about 1 to 2 litres of urine per day.

The volume is influenced by:

1. Fluid intake and production of 'metabolic water' in the body (water produced by oxidation of foodstuff)
2. Water loss through extrarenal routes like the skin, lungs and GI tract
3. Temperature and humidity in the atmosphere

Polyuria – Increased urinary volume (>2 L/day)
Oliguria – Decreased urinary volume (<400 mL/day)
Anuria – Cessation of urine output (urine excretion is blocked)

However, total cessation of urine production is not possible despite there being no water intake. This is because a minimum quantity of water (300–500 mL/day) is required as the medium to eliminate the toxic and waste products (formed during metabolism) from the body.

Polyuria is seen after excess fluid intake, ingesting spicy foods and drinking beverages like beer, coffee and tea (due to their diuretic effect). Deficiency of antidiuretic hormone (ADH) causes **diabetes insipidus** characterised by polyuria. In **diabetes mellitus**, increased concentration of glucose blocks reabsorption of water in the renal tubules, causing polyuria. Polyuria may also be seen due to tubular dysfunction in **nephrogenic diabetes insipidus**. Drugs like diuretics also cause polyuria.

High urine output at night (**nocturia**) is one of the early symptoms of chronic renal disease. **Oliguria** is seen in diarrhea, vomiting, excessive sweating (fluid removal through extrarenal routes), in fever due to less fluid intake and renal diseases (for example, acute glomerulonephritis). **Anuria** may be due to bilateral obstruction of the ureters, resulting in total urinary obstruction. Shock, mercury poisoning and incompatible blood transfusion cause anuria.

Appearance

Normally it is clear. Excess phosphate excretion or excretion of pus cells due to infections makes it turbid. When lymph is excreted (*chyluria*), it appears milky white. Urine becomes turbid on standing due to precipitation of phosphate.

Colour

Normal urine is pale yellow in colour. The pigment responsible for this is urochrome. In jaundice, urine appears greenish yellow. In '**alkaptonuria**' (an inborn error of tyrosine metabolism), the urine becomes smoky or black due to the presence of homogentisic acid. Change in colour is also seen in a number of disorders, for example, hemoglobinopathies (red), due to the presence of abnormal bile pigments (yellow), melanins (black) and porphyrins (red-brown).

Odour

Normal urine is pungent (ammoniacal) in odour.

Fruity odour	–	Due to ketonuria (excretion of ketone bodies in severe diabetes mellitus)
Foul smell	–	Bacterial infection
Cabbage odour	–	In tyrosinemia
Mousy odour	–	In phenylketonuria

pH

Test the reaction of urine with litmus paper. Urine is usually acidic. The normal pH of urine ranges between 4.8 and 7.5, mean pH 6.0. On standing, urea will be converted to ammonia by bacterial action and it will increase the pH. Diet also influences the pH of urine. Ingestion of protein-rich diet increases urinary acidity. Diet rich in vegetables and fruits decreases urinary acidity. The urine excreted after meals is alkaline and is known as **alkaline tide**. Usually, the citric acid and tartaric acid present in fruits and vegetables are converted to bicarbonates and make the urine alkaline. The rise in pH as a result is compensated by the secretion of HCl in the gastric juice. In **diabetic ketoacidosis**, ketonuria occurs, causing excretion of acidified urine because of the presence of ketone bodies (acetoacetic acid, β-hydroxybutyric acid).

Specific gravity

Specific gravity is used to determine the mass (weight) of solutions. It relates the weight of 1 mL of the solution and the weight of an equal volume of pure water at 4°C.

Urine has higher specific gravity than that of water because of the various solutes present in it. The normal range of specific gravity of urine is between 1.010 and 1.030 (usally 1.015 to 1.025). A higher specific gravity indicates a more concentrated urine.

Specific gravity is measured with an instrument known as the **urinometer**. It consists of a weighted cylinder and a thin stem calibrated from 1.000 to 1.060 (representing 1.000 to

1.060 degrees of specific gravity) (Fig. 3.1). The instrument is usually calibrated for use at a temperature of 15°C or 20°C.

How to measure the specific gravity of urine?

The urine is poured into a wide-mouthed vessel, and the urinometer is put into it so that the weighted cylinder sinks in the urine, leaving the graduated stem floating. Ensure that the urinometer does not touch the side walls of the vessel in which it is floating. Observe the graduation which lies level with the surface of urine and note the reading. Also note the temperature of urine.

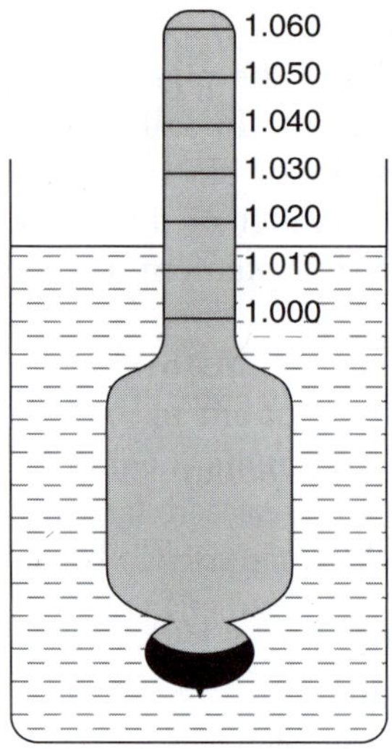

Fig. 3.1 *Urinometer*

Calculation

Suppose the meniscus of the urine coincides with the reading 1.015; it corresponds to the specific gravity 1.015. But the urinometer is calibrated for use at a temperature of 15°C. Hence, the specific gravity should be corrected to room temperature which may be above or below 15°C at which the urinometer was calibrated.

Temperature correction: 0.001 is to be added for every 3°C above and 0.001 is to be deducted for every 3°C below in the temperature at which the urinometer is calibrated (15° C). Suppose the reading coincides with 1.015, it will be corrected to room temperature as follows. The difference between room temperature 37° and 15° (calibrated temperature) is 22°C. This, when divided by 3 gives 7. So the corrected specific gravity = 1.015 + (7 × 0.001) = 1.022.

Clinical significance

The specific gravity is inversely proportional to the volume of urine excreted. Thus in **diabetes insipidus** (which is characterised by polyuria due to ADH deficiency), the specific gravity is as low as 1.001, whereas in febrile conditions a smaller volume of urine is excreted, having high specific gravity.

In contrast, in **diabetes mellitus**, despite polyuria, specific gravity is high because of the high amount of glucose present.

Microscopic Examination

Microscopy is an integral part of routine urinalysis in screening for disease. Microscopic examination is usually done for cells, casts, crystals and bacteria. It is carried out in microbiology and pathology laboratories.

Chemical Analysis

Urine contains both inorganic and organic constituents.

Inorganic constituents

Na^+, K^+, Ca^{2+}, Mg^{2+}, Cl^-, $H_2PO_4^-$, HPO_4^{2-}, SO_4^{2-}, HCO_3^-, etc.

Organic constituents

Urea, uric acid, creatinine (total quantity of the NPN excreted is 10–20 g/day), certain detoxified products, for example, ethereal sulphates.

Composition of normal urine

Volume	:	1000–2000 mL/day
pH	:	4.8–7.5
Specific gravity	:	1.015–1.025
Albumin	:	Traces
Sugar	:	Nil

Inorganic constituents

Calcium	:	0.2 g/day
Chlorides	:	6–16 g/day
Sulphates	:	0.5–1.5 g/day
Phosphates	:	0.8–1.5 g/day
Sodium	:	3–5 g/day
Potassium	:	2–4 g/day

Organic constituents

Ammonia	:	0.5–0.8 g/day
Urea	:	25–30 g/day
Uric acid	:	0.5–1.0 g/day
Creatinine	:	1.0–2.0 g/day

Tests for Inorganic Constituents

Chlorides

1. To 1 mL of urine, add 0.5 mL of concentrated HNO_3 and 1 mL of $AgNO_3$.
2. Chloride is precipitated as silver chloride, forming a curdy white precipitate.

Clinical significance

Decrease in urinary chlorides is seen in edematous conditions, diabetes insipidus and pneumonia. Increase of urinary chlorides is seen in Addison's disease and use of diuretics.

Sulphates

1. To 2 mL of urine, add 2 mL of barium chloride solution.
2. Sulphate is precipitated as barium sulphate, forming a white precipitate.

Clinical significance

Sulphates are derived from sulphur-containing amino acids. Urinary content of inorganic sulphates decreases in conditions related to impaired renal function and increases in high protein intake and increased tissue protein breakdown.

Phosphates

1. To 5 mL of urine, add a few drops of concentrated HNO_3 and a pinch of ammonium molybdate and heat.
2. A canary-yellow–coloured precipitate of ammonium phosphomolybdate is observed.

Clinical significance

Urinary PO_4 is derived from the breakdown of phospholipids, nucleotides and phosphoproteins. Urinary content of phosphates decreases in diarrhea, nephritis, hypoparathyroidism and pregnancy and increases in rickets, osteomalacia, hyperparathyroidism and various acidotic conditions.

Calcium

Normally, most of the Ca and Mg are excreted in feces.

Test for urinary calcium (Sulkowitch's test)

1. To 5 mL of urine, add 5 drops of 1% acetic acid and 5 mL of potassium oxalate solution.
2. Observe the formation of a white precipitate of calcium oxalate.

Clinical significance

The vast majority of **kidney stones** (up to 75%) contain calcium oxalate and calcium phosphate crystals. Magnesium ammonium phosphate (**struvite**) stones account for about 15% of cases. In such cases, estimating calcium levels in urine helps in evaluating patients suffering from **renal calculi** (kidney stones).

An increase in the urinary content of calcium is seen in renal stones, hyperparathyroidism, hypervitaminosis 'D', multiple myeloma and renal tubular acidosis. Certain drugs like diuretics and steroids may enhance urinary Ca excretion.

Tests for Organic Constituents

Organic sulphate (ethereal sulphate)

Ethereal sulphates represent detoxified products of phenols and are formed in the liver. About 100 mg of organic sulphates are excreted per day.

1. To 5 mL of urine, add 2 mL of barium chloride solution and 2 mL of concentrated HCl. Mix well and filter.
2. Divide the filtrate into two portions. Boil one and compare with the control.
3. Turbidity indicates the presence of ethereal sulphates. Concentrated HCl hydrolyses ethereal sulphate to inorganic sulphate, which then gives a precipitate with $BaCl_2$.

Urobilinogen

Urobilinogen, derived from bilirubin by the action of bacterial flora in the intestine, enters into circulation and is excreted by the kidneys.

1. To 5 mL of urine, add 1 mL of **Ehrlich's reagent**, mix and stand for 5 minutes. (Ehrlich's reagent: 100 ml of water + 0.7 g of paradimethyl amino benzaldehyde + 15 mL of concentrated HCl).
2. A red colour is seen when viewed through the mouth of the test tube. Urobilinogen reacts with para dimethyl amino bezaldehyde of the reagent to form a red colour.

Clinical significance

Normal urine contains traces of urobilinogen and hence the test is only weakly positive. Upon standing, urobilinogen in urine is oxidised to urobilin. Urobilin does not answer the above test.

Absence of urobilinogen in urine is suggestive of obstructive jaundice. Excess of urobilinogen in urine is indicative of hemolytic jaundice.

Non-protein Nitrogenous (NPN) Substances in Urine

Ammonia

Urinary ammonia is derived from glutamine and other amino acids in the kidneys. Acid forming foods increase ammonia excretion.

1. To 5 mL of urine, add 1 drop of phenolphthalein.
2. Then add 0.1 N NaOH, drop by drop, till a pink colour is seen.
3. Now, heat the solution and hold a glass rod dipped in phenolphthalein at the mouth of the test tube. The tip of the glass rod turns pink for a moment, representing the liberation of ammonia.
4. Phenolphthalein gives a pink colour in alkaline media.

Urea

Urea is the end product of protein metabolism. It is synthesised in the liver and its excretion depends on the protein content in diet. About 20–30 g of urea is excreted per day.

Tests for urea

1. Sodium hypobromite test
2. Specific urease test

For more details on these tests, please see Chapter 5.

Clinical significance

Increase: The urinary content of urea is elevated whenever protein intake and catabolism are increased, for example, fevers, diabetes mellitus and hyperfunction of the adrenal cortex.
Decrease: Cirrhosis of the liver, nephritis and severe acidosis.

Uric acid

Uric acid is the end product of purine catabolism. Uric acid is excreted at 0.5–1.0 g/day.

Tests for uric acid

1. Phosphotungstic acid test
2. Schiff's test

3. Murexide test

For more details on these tests, please see Chapter 5.

Clinical significance

Hyperuricemia (elevated uric acid levels) is mostly associated with **gout**. It is also seen in leukemias and Wilson's disease. Accumulation of uric acid crystals in the synovial fluid causes inflammatory changes, resulting in gouty arthritis. Sodium urate crystals are deposited in the form of **tophi** in peripheral joints, making them extremely painful.

Uric acid crystals are also deposited in the collecting tubules and the lower urinary tract, resulting in calculi (stones) and causing renal damage.

Creatinine

The quantity of creatinine excreted by normal adults is 1.0–2.0 g/day. Creatinine is derived from muscle phosphocreatine and its levels are not dependent on protein intake. Hence its estimation in serum is considered as a more reliable investigation in exploring renal disease.

Test for creatinine

Jaffe's test: Jaffe's reaction is the specific test for creatinine. *For more details on this text, please see Chapter 5.*

Clinical significance

The amount of creatinine excreted is proportional to the muscle phosphocreatine content. The daily excretion of creatinine is more or less constant when compared to that of other NPN substances. Increased creatinine excretion is seen in muscular dystrophy, hyperthyroidism, diabetes mellitus and starvation.

Viva Questions

1. What is the volume of normal urine output per day? Describe the terms *polyuria, oliguria* and *anuria.*
2. What are the common causes of turbidity in a fresh urine sample?
3. What is the pigment responsible for the normal colour of urine?
4. What is alkaline tide?
5. How is the specific gravity of urine measured? Name one condition where the specific gravity of urine is high.
6. What is 'temperature correction' while reading the urinometer?
7. Despite polyuria, specific gravity is high in diabetes mellitus. Why?
8. Name the conditions in which large amounts of calcium are excreted in urine.
9. Name the NPN substances excreted in urine.
10. Is urobilinogen normally present in urine? Name one condition where excretion of urobilinogen is increased. Name the test to identify urobilinogen.
11. What are the compositions of various kidney stones?
12. Which substance constitutes the major portion of NPN?
13. What are ethereal sulphates?
14. Name the conditions in which excretion of urea in urine is increased.
15. When is the excretion of creatinine increased?

4 Analysis of Abnormal Constituents of Urine

BI 11.4; 11.20

Substances that occur in extremely low concentrations and hence are not detectable in the urine of normal healthy individuals but which are present in urine in detectable amounts (high concentrations) in certain diseased conditions are said to be **abnormal** or **pathological constituents** of urine.

Usually, the analysis is done on a properly preserved 24-hour urine sample. When this is not possible, the urine voided in the morning can be used. Nowadays, reagent strips (**dipsticks**) have replaced many of the routine tests in urine analysis. Reagent strips for glucose, ketone bodies, proteins, hemoglobin and pH are commercially available and are simple to use.

Before detecting the abnormal constituents of urine, note the physical characteristics.

Physical Characteristics

Volume

Polyuria is seen in diabetes mellitus, diabetes insipidus and after using certain drugs like salicylates, diuretics and digitalis. Oliguria is observed in diarrhea, vomiting, fever and nephritis. Total cessation of urine output (*anuria*) may occur during shock, acute nephritis, incompatible blood transfusion and bilateral renal stone formation.

Colour

Urine appears yellow when bilirubin is present. It becomes red or smoky brown when blood is present. Presence of pus makes it milky, while melanin turns urine black.

pH

Acidic urine is passed in fever and diabetes mellitus. Urinary retention makes it alkaline.

Specific gravity

It is increased in diabetes mellitus, fever and acute nephritis and decreased in diabetes insipidus.

Chemical Constituents

The abnormal constituents which occur in urine in various pathological conditions include glucose, ketone bodies, proteins, bile salts, bile pigments and blood. Nowadays, the chemical analysis of urine specimens is carried out using commercially available disposable strips.

1. Glucose

Glucose or any other sugar is not excreted by healthy individuals. Glucose is present in glomerular filtrate in the same concentration as in blood plasma. But it is completely reabsorbed by renal tubular epithelial cells. The rate of glucose absorption is expressed as '**tubular maximum for glucose**' (**TmG**), which is 350 mg/minute.

When the blood levels of glucose are elevated to such a level that glucose cannot be reabsorbed any more by the renal tubular cells and hence appears in urine, that level is known as the '**renal threshold for glucose**', which is 180 mg/100 mL. Excretion of glucose or any other sugar in urine is called '**glycosuria**'. Glycosuria occurs mainly in **diabetes mellitus**. Benedict's test is routinely used to detect the sugar levels in urine.

Benedict's Test

It is the test used to detect glucose and other reducing sugars in the urine samples of patients suspected to be suffering from hyperglycemic conditions such as diabetes mellitus.

Procedure

1. To 5 mL of Benedict's reagent, add 8 drops of urine.
2. Boil for 2 minutes over a flame.

Observation

A light green, yellow, orange or brick red precipitate is produced depending on the urinary content of the reducing sugar.

Inference

A positive Benedict's test indicates the presence of glucose or any other reducing sugar in the given urine sample. Benedict's test is considered to be a '**semi-quantitative test**' as the colour of the precipitate roughly corresponds to the concentration of the reducing sugar.

Interpretation of the result of Benedict's test

Blue colour	No sugar in urine
Green precipitate	[(+) one plus] concentration of sugar is 0.5 g/dL
Yellow precipitate	[(++) two plus] 1 g/dL
Orange precipitate	[(+++) three plus] 1.5 g/dL
Brick red precipitate	[(++++) four plus] 2 g/dL and more

Glycosuria

The pathological basis of glycosuria has two mechanisms.

1. When the blood glucose level exceeds the renal threshold point (180 mg/dL), large quantities of glucose exceeding TmG enter into tubules and are excreted in urine. This is known as '**hyperglycemic glycosuria**'. **Diabetes mellitus** is the commonest cause of hyperglycemic glycosuria.
2. Any decrease in the glucose reabsorption capacity of the renal tubular cells results in glucose excretion into urine. This occurs even when the glucose concentration in blood is well within the normal limits. This is known as '**renal glycosuria**'.

Hyperglycemic glycosuria

Glycosuria in diabetes: Diabetes mellitus is the most common condition in which glycosuria is seen due to hyperglycemia. The sugar present in urine is always glucose. However, the diagnosis of diabetes should not be made based only on glycosuria because glycosuria may occur even during normoglycemia (renal glycosuria). The diagnosis of DM should always be based on a corresponding estimation of glucose levels in blood.

Benedict's test to detect glucose in urine (glycosuria) is most commonly done to screen for diabetes.

Double void technique: One should look out for glucose in a specimen of urine produced by the kidneys at that time. The patient is asked to empty the bladder and discard the specimen. The next specimen passed 10–15 minutes later is collected and tested. The double void technique ensures that the specimen being tested reflects the blood glucose at that time and has not been stored in the bladder for some time, before collection.

The urine specimen collected along with a postprandial blood sample (two hours after a meal) is more valuable in detecting diabetes than a fasting specimen.

Can the diagnosis of diabetes be made based on glycosuria alone?

Glucose in urine cannot be the sole diagnostic criterion for diabetes because of the individual variations in renal threshold. Glycosuria due to a low renal threshold for glucose, which is seen during pregnancy and in young persons, is a more frequent cause of glycosuria than diabetes. Renal glycosuria is a benign condition totally unrelated to diabetes. Hence when glycosuria is detected, it is always wise to estimate the glucose concentration in blood. Only unambiguous hyperglycemia confirms diabetes.

Alimentary glycosuria

When a carbohydrate-rich diet is ingested, blood glucose rises and may sometimes cross the renal threshold and produce glycosuria. This condition is known as alimentary glycosuria. Persons with impaired glucose utilisation may exhibit such type of hyperglycemic glycosuria and should be kept under observation and preferably be screened for diabetes periodically.

Transient glycosuria

It is seen in persons undergoing emotional stress. Excessive secretion of catecholamine hormones produces hyperglycemia and thus results in glycosuria.

Lactosuria

Lactose is present in the urine of normal women during the **third trimester of pregnancy** and during **lactation**.

Galactosuria

It is due to deficiency of the enzyme galactose-1-phosphate uridyl transferase.

Fructosuria

It is seen in **hereditary fructose intolerance** or fructokinase deficiency.

Pentosuria

It is seen in **essential pentosuria** due to the deficiency of the enzyme xylitol dehydrogenase.

Hyperglycemic glycosuria may also be seen in hyperthyroidism, hyperpituitarism and Cushing syndrome.

Renal glycosuria

Glucose is excreted in urine due to lowering of the renal threshold. It may happen due to renal tubular transport defects, for example, **Fanconi syndrome**.

Renal threshold may also be lowered during pregnancy, resulting in glycosuria (**physiological glycosuria**). But pregnancy may be associated with diabetes (**gestational diabetes mellitus – GDM**) causing glycosuria. These two conditions can be differentiated by estimating blood sugar levels. *It should be noted that renal glycosuria is more common than glycosuria due to diabetes.* In certain diseases affecting renal tubular function and in heavy metal poisoning, renal glycosuria occurs.

To summarise, apart from glucose (in diabetes mellitus) the following substances in urine produce a positive reducing test. Lactose (pregnancy and lactation), galactose (galactosemia), fructose (hereditary fructose intolerance), pentoses (essential pentosuria), glucuronide derivatives of drugs, homogentisic acid (alkaptonuria), creatinine (in high concentrations), ascorbic acid (in high vitamin C intake) and salicylic acid (in aspirin overdose).

2. Ketone Bodies

Acetoacetic acid, β-hydroxy butyric acid and acetone which are derived from the metabolism of fatty acids are known as '**ketone bodies**'. When the concentration of ketone bodies goes beyond their renal threshold (3 mg/dL), they appear in urine. **Ketonuria** usually occurs in **severe uncontrolled diabetes mellitus and starvation.**

Rothera's Test

Procedure

1. Take 10 mL of urine, add a pinch of solid ammonium sulphate, a little at a time, and shake till the solution becomes saturated.
2. Add 2–3 drops of freshly prepared sodium nitroprusside and mix gently.
3. Add 1 mL of strong ammonium hydroxide along the side walls of the test tube.

Observation

A purple-coloured ring is obtained at the junction of the two liquids.

Inference

The ketone bodies react with sodium nitroprusside in the presence of NH_3 and form a purple-coloured ring. A positive test is indicative of the presence of acetoacetic acid and acetone.

Gerhardt's Test

Procedure

1. To 3 mL of urine, add ferric chloride, drop by drop (filter if a precipitate of ferric phosphate is seen).

Observation

A port wine colour is obtained.

Inference

A positive test is given by acetoacetic acid and not by acetone. It is less sensitive than Rothera's test. False positive Gerhardt's test may be obtained by salicylic acid and salicylates. This can be differentiated by heating. A urine sample which produces a positive test before boiling and a negative test after boiling is indicative of the presence of acetoacetic acid.

Iodoform Test

Acetone can give iodoform when treated with alkaline iodine solution.

Test for β-Hydroxy Butyric Acid

All of the reactions mentioned above do not detect β-hydroxy butyric acid. There is no specific test for detection, but it can be detected after oxidising to acetoacetate.

Procedure

1. Add a few drops of acetic acid to 3 mL of urine diluted 1:1 with distilled water.
2. Boil for a few minutes to remove acetoacetate and acetone.
3. Add 1 mL of H_2O_2 and heat gently. Cool and perform Rothera's test. A positive test is indicative of β-OH butyric acid.

Black's reaction: β-OH butyric acid when oxidised with H_2O_2 and treated with a mixture of ferric and ferrous chorides produces a rose colour.

Clinical significance

Ketone bodies are synthesised in the liver but they are not utilised by it. Ketone bodies are excreted into urine *(ketonuria)* in **severe uncontrolled diabetes mellitus** and **starvation**. Ketone bodies may reach as high as 100 mg/dL in blood and urinary excretion may be as high as 500 mg/day in severe uncontrolled diabetes mellitus.

Ketoacidosis

Ketone bodies, acetoacetate and β-hydroxy butyrate are acids. These ketoacids are normally buffered in tissues and blood. But, because of their excretion in large quantities, there is some loss of buffer cations which also depletes the alkali reserve causing **ketoacidosis**. Patients of diabetic ketoacidosis (DKA) will have a typical breathing pattern known as '**Kussmaul's respiration**' (a compensatory hyperventilation). The breath of patients of DKA smells sweeter because of the presence of acetone.

Biochemical mechanisms underlying ketosis

Impaired glucose metabolism and increased lipolysis occur in diabetes mellitus due to insulin deficiency. Lack of insulin hampers the entry of glucose into cells and hence the body depends on fat for energy. This results in the production of large quantities of acetyl CoA, which in turn is used in the synthesis of ketone bodies, elevating their levels in blood *(ketosis)*. In **starvation**, the body meets its energy requirements by utilising the stored fats. The oxidation of fatty acids results in excess production of acetyl CoA which cannot be fully handled by the citric acid cycle. Furthermore, the Krebs cycle is impaired due to the deficiency of oxaloacetate, since most of it is diverted for glucose formation (to meet the energy requirements of tissues such as the brain). This causes accumulation of acetyl CoA and its conversion to ketone bodies, resulting in ketosis.

3. Proteins

Proteins excreted in normal urine are in traces (less than 30 mg per day). Presence of proteins in urine in detectable quantities is known as '**proteinuria**' and is always pathological. Proteinuria is due to a glomerular leak or the inability of the tubules to reabsorb protein completely. As the major chunk of protein excreted in urine is albumin, it is commonly called '**albuminuria**'. Albuminuria is usually associated with renal disease.

Tamm–Horsfall protein is a small glycoprotein that occurs in normal urine and is secreted by the mucous glands of the genitourinary tract (its function is protection of the epithelial cells of the tract). But usually it is not detectable by the routine tests employed in urine analysis because of its extremely low concentration.

Heat and Acetic Acid Test

Procedure

1. Take a test tube that is two-thirds full of the urine sample. Add a few drops of 1% acetic acid. Do not mix.
2. Hold the tube over a flame in a slanting position and boil the upper portions of the tube, leaving the lower part as control.

Observation

A white turbidity/precipitate in the heated portion indicates the coagulum of albumin. An unchanged lower portion serves as control.

Inference

When heated and treated with acetic acid, the albumin present in urine becomes coagulated, as the pH is close to its isoelectric point. If there is no turbidity, albumin may be absent in urine.

Sulphosalicylic Acid Test

1. Take 2 mL of urine in a test tube and add a few drops of 20% sulphosalicyclic acid. Albumin, if present, will appear as a white precipitate.

Heller's Nitric Acid Test

1. Take 2 mL of concentrated HNO_3 in a test tube and layer the urine cautiously on to it, so as to obtain a sharp line of demarcation. Proteins, when present, produce a white ring at the junction of the two fluids.

Causes of proteinuria

Proteinuria may be seen after strenuous exercise and during pregnancy (physiological). Fever, abdominal tumours and cardiac disease may produce proteinuria by affecting the circulation of the kidneys. **Proteinuria is always associated with renal pathology**. Acute glomerulonephritis, early stages of chronic glomerulonephritis, nephrotic syndrome, nephrosclerosis and carcinoma of the kidneys show proteinuria.

Biochemical mechanisms underlying proteinuria

There are five underlying mechanisms for proteinuria: (1) increased glomerular permeability, (2) defective tubular reabsorption (3) increased production of certain low-molecular-weight abnormal proteins such as the light chains of immunoglobulins, (4) abnormal secretion of protein into the urinary tract (due to inflammation or malignancy) and (5) hemorrhage.

The renal glomerulus acts as a molecular sieve excreting any substance inversely proportional to its molecular radius. Since albumin is small and globular, large amounts of albumin filter into the glomerular filtrate despite its acidic charge at normal pH. It can be noted that filtration is increased if acidic groups are masked (as occurs by glycosylation in diabetes mellitus).

Most of the filtered albumin is usually reabsorbed by the proximal tubules, limiting the excretion of albumin in urine to minimal levels (<30 mg/day). Excretion above this indicates, as mentioned above, glomerular filtration that surpasses the reabsorption capacity of the proximal tubules (as seen in **nephrotic syndrome**), proximal tubular damage, hematuria or a combination of these. Increased glomerular filtration also occurs during strenuous physical exercise and prolonged fever (the excretion of proteins may be up to 300 mg/day in the urine of healthy individuals after exercise).

In nephrotic syndrome, even though the glomerular filtration of all proteins increases, only the levels of albumin are decreased in plasma compared to other proteins. This is because glomerular sieving retains the bigger proteins while smaller proteins are lost in urine (including albumin). The loss of proteins in urine is compensated by an increase in the production of different proteins (including albumin) by the liver. As a net result of increased production, and retention by sieving, the concentrations of proteins with high molecular mass such as α_2 macroglobulin, larger isotypes of haptoglobins and lipoproteins outnumber the concentrations of smaller proteins such is albumin, leading to **reversal of the A/G ratio**. The electrophoretic separation of plasma proteins shows a marked elevation in α_2 globulins and lipoproteins and a drastic reduction in most other proteins including albumin (*see Fig. 21.2*).

Bence–Jones proteins

Bence–Jones proteins appear in urine samples of patients of **multiple myeloma** (a plasma cell tumour). They are the light chains of immunoglobulins produced in abnormally high quantities which are excreted into urine.

Microalbuminuria

Albumin occurs in normal urine in traces (2.5–30 mg/day). When its concentration in urine lies between 30 and 300 mg/day, the condition is commonly known as '**microalbuminuria**' (more correctly termed **minimal albuminuria or paucialbuminuria**). It occurs as a complication of diabetes and hypertension. When albuminuria is >300 mg/day, it is called **macroalbuminuria** and is indicative of overt **renal disease** (diabetic nephropathy). From the above discussion, it is evident that microalbuminuria is an intermediary stage between normal health and overt renal pathology and hence it can be taken as a warning sign of **future renal failure**. Increased urinary albumin excretion (paucialbuminuria) is highly predictive of **diabetic nephropathy, end stage renal disease and proliferative retinopathy**. Patients of type I diabetes mellitus are at high risk of developing all these complications if left untreated.

Tests to detect Bence–Jones proteins

1. **Heat test**: When the urine is heated to 50–60°C, Bence–Jones proteins if present, are precipitated. On further heating, they re-dissolve. They precipitate again on cooling.
2. **Bradshaw's test**: When urine is layered on concentrated. HCl, a white ring of precipitate is formed, indicating the presence of Bence–Jones proteins.
3. **Electrophoresis**: Bence–Jones proteins are best detected by zone electrophoresis and immuno electrophoresis of concentrated urine.

Post-renal proteinuria

Post-renal proteinuria refers to the excretion of proteins that arise from the urinary tract below the kidneys. It usually occurs in inflammation and malignancy. Microscopic examination of urinary sediment is carried out for inflammatory cells and malignant cells. The detection of RBCs and WBCs in casts of centrifuged urinary sediment is diagnostic proof of their origin from the kidneys.

4. Blood

Passing of whole blood including RBCs in urine is called '**hematuria**'. Urine appears smoky brown when blood is present. The benzidine test is performed to detect hematuria.

Benzidine Test

Procedure

1. Take 3 mL of a saturated solution of benzidine in glacial acetic acid and add 1 mL of H_2O_2 to it. Mix well.
2. Add an equal quantity of urine sample.

Observation

A blue or green colour develops, which is stable only for a few minutes and then changes to brown.

Inference

A positive test is indicative of blood in urine.

Mechanism of benzidine reaction

The test depends on the peroxidase activity of Hb. Oxygen liberated from H_2O_2 by peroxidase activity is accepted by benzidine to produce the blue colour compound.

The test is also positive when pus cells are present. Pus cells possess peroxidase activity and hence give a positive reaction. To avoid this, the fluid should be heated first before testing. With blood, the test is positive even after heating because heme (which catalyses the reaction) is heat stable.

Caution

☞ Benzidine is carcinogenic and hence one should be careful while performing the test. Because of its carcinogenic nature, benzidine is now replaced by O-toluidine (1.2% solution of O-toluidine can be used).

Nowadays, hemastix papers are available for easy and rapid testing of blood in urine.

Guaiac Test

Procedure

1. Dispense 5 mL of fresh urine into a heat-resistant glass tube.
2. Add freshly made gum guaiac reagent until turbidity forms.
3. Add about 10 drops of freshly prepared H_2O_2.
4. Look for blue colour. If a blue colour develops, boil the urine in a container with water for 15–20 minutes.

Observation

- Blue colour remains after boiling – positive test
- Blue colour disappears after boiling – negative test

Inference

A positive test indicates the presence of hemoglobin in urine (hemoglobinuria).

Causes of hematuria

Hematuria suggests bleeding in the renal tract. It occurs because of bleeding in the urinary tract due to **trauma, infection, enlargement of prostate** causing rupture of engorged venous plexus, **bladder cancer, acute glomerulonephritis** and **renal calculi** (**stones**). Administration of certain drugs like salicylates, sulphonamides and barbiturates may lead to hematuria. Physiologically, hematuria may be seen after strenuous exercise. Sometimes the urine may be contaminated by menstruation, resulting in hematuria.

Excretion of free Hb followed by hemolysis is known as '**hemoglobinuria**'. This occurs in severe burns, hemolytic jaundice, incompatible blood transfusion and poisoning. The detection of WBCs in the urine suggests *acute inflammation* and possible *urinary tract infection*. Bleeding may be associated with pain in the loins in **loin pain hematuria syndrome**. Recurrent episodes of hematuria in combination with respiratory infections are seen in **IgA nephropathy**.

Certain other causes of red or dark urine may sometimes be confused with hematuria. They include myoglobinuria, alkaptonuria, porphyria and administration of certain drugs (rifampicin and levodopa).

5. Bile salts

Sodium taurocholate and sodium glycocholate are the common bile salts. They are strong emulsifying agents.

Hay's Test

Procedure

1. Take two test tubes and label them 'T' (test) and 'C' (control).
2. In the 'T' test tube, take 5 mL of urine, and in the 'C' test tube, take 5 mL of water.
3. Sprinkle sulphur powder gently in both the tubes.

Observation

Sulphur powder sinks to the bottom in test whereas it floats in control.

Inference

A positive test is specific for bile salts. The presence of bile salts lowers the surface tension of urine and hence sulphur sinks to the bottom.

Pettenkoffer's Test

1. Take 5 mL of urine in a test tube and add 1 drop of glucose solution.
2. Mix and then add 5 mL of concentrated H_2SO_4 along the side walls of the tube.
3. A purple ring at the junction of the two fluids indicates the presence of bile salts. The furfural formed reacts with the bile salts, producing a purple ring.

Oliver's Test

1. Take 5 mL of urine in a test tube and add an equal quantity of 1% acidified peptone solution.
2. The appearance of white milkiness or precipitate is indicative of the presence of bile salts.

6. Bile pigments

Bilirubin occurs in two forms, conjugated and unconjugated. Only the conjugated form is soluble in water and hence it is the conjugated bilirubin that is excreted in urine. Bilirubinuria is invariably pathological. Conjugated bilirubin from the liver enters into the GI tract through the biliary canal and is degraded further. However, some of it re-enters the liver through the portal vein and is re-secreted in bile *(entero-hepatic circulation)*. Any interruption to this constant flow due to mechanical obstruction causes leaking of conjugated bilirubin into blood, some of which is excreted into urine.

Bilirubin, biliverdin and bilicyanin are the common bile pigments. The following tests are used to detect them in urine.

Fouchet's Test

Procedure

1. Take 10 mL of urine, add a pinch of magnesium sulphate and boil.
2. While boiling, add 5 mL of 10% $BaCl_2$ solution, drop by drop, till the maximum precipitate is attained. This precipitate of barium sulphate absorbs bile pigments and thus turns yellow.
3. Filter and discard the filtrate.
4. Unfold the filter paper on a dry filter paper and press lightly to squeeze out moisture.
5. To the precipitate, add 2 drops of Fouchet's reagent (ferric chloride in trichloro acetic acid).

Observation

Note the development of green or blue colour.

Inference

Ferric chloride oxidises bilirubin to biliverdin (green) and bilicyanin (blue).

Gmelin's Test

1. Layer some urine carefully on 2 mL of concentrated HNO_3 in a test tube. Green, blue or violet rings appearing at the junction of the liquids indicate the presence of bilirubin in urine.

Clinical significance of bile salts and bile pigments in urine

Bile salts and bile pigments are excreted in urine in **obstructive jaundice**.

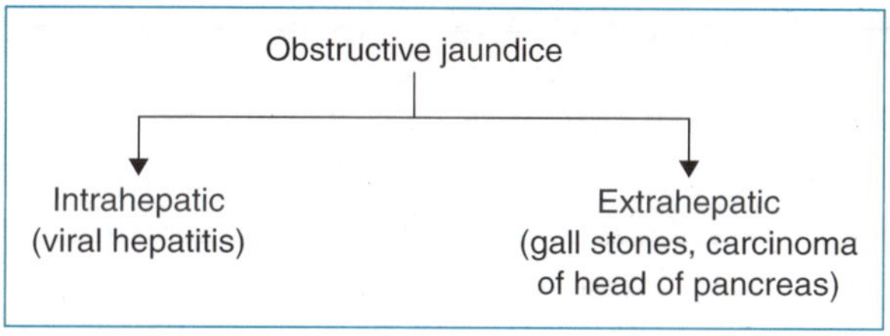

In the case of jaundice, urine should be examined for the presence of bilirubin, bile salts and urobilinogen. Only conjugated bilirubin is soluble in water and is excreted in urine. Hence, in prehepatic jaundice, as the unconjugated bilirubin is increased in blood, it does not appear in urine.

In obstructive jaundice, bilirubin is conjugated and is not excreted by the normal passage due to the obstruction, and so it is regurgitated into the bloodstream. This is then excreted through urine. In cases of obstruction, bile does not reach the intestine and so urobilinogen may be decreased or absent in urine.

In hepatocellular jaundice, urobilinogen is initially elevated but decreases or disappears later when there is obstruction, and reappears when the obstruction is cleared. The detection of urinary bilirubin and urobilinogen by simple tests is therefore useful in following the course of a case of jaundice.

Urobilinogen is absent in urine when there is obstruction to bile flow. The first indication of the recovery is the reappearance of urobilinogen in urine. It is increased in urine in hemolytic anemias.

Urobilinogen

Normal urine contains little urobilinogen. Upon standing, urobilinogen in urine is oxidised to urobilin, deepening the colour of urine.

Ehrlich's test

Procedure

1. To 10 mL of urine, add 1 mL of Ehrlich's reagent (paradimethyl amino benzaldehyde in HCl).

Observation
A red colour develops.

Inference
Indicates the presence of urobilinogen. Urobilinogen reacts with paradimethyl amino benzaldehyde of the reagent to form a red-coloured complex.

Schlesinger's test
1. To 5 mL of urine, add 2 drops of N/100 iodine.
2. Mix well with 5 mL of 10% alcoholic solution of zinc acetate.
3. Filter. The filtrate shows green fluorescence, indicative of urobilinogen.

Clinical significance
Normal urine contains traces of urobilinogen and hence the tests are weakly positive. But in **hemolytic jaundice**, the excretion of urobilinogen is increased and the tests are strongly positive. Urobilinogen is totally absent or decreased in **obstructive jaundice**.

Summary of urinary findings in different types of jaundice

Type of jaundice	Bile salts	Bile pigments	Urobilinogen
1. Hemolytic jaundice	Absent	Absent	++
2. Hepatocellular jaundice	+	++	Normal or decreased
3. Obstructive jaundice	++	+++	Absent or decreased

Test for Nitrite
When bacteria is present in urine, they convert the nitrate (from the diet) to nitrite. A dipstick test is performed in this regard. A positive test indicates a urinary tract infection.

Viva Questions
1. What are the abnormal or pathological constituents of urine?
2. What is glycosuria? What are the conditions in which it occurs?
3. Name the test to identify reducing sugars in urine.
4. What is renal glycosuria? What is the renal threshold for glucose?
5. How do we differentiate pathological glycosuria from the physiological? Can glycosuria occur when blood glucose levels are normal?
6. What are the tests used to identify ketonuria? Does Rothera's test identify β-hydroxy butyric acid?
7. Under which disease conditions does ketonuria occur? In which condition are both glucose and ketone bodies excreted in urine?
8. Which protein is commonly excreted in the early stages of renal disease?
9. What are Tamm–Horsfall proteins? Are these normally found in urine?
10. Why are proteins not normally excreted in urine?
11. In which diseases does proteinuria occurs?
12. What are the underlying biochemical mechanisms of proteinuria?

13. What is microalbuminuria?
14. What are Bence–Jones proteins? How are they detected in urine?
15. What is hematuria? Name the test that detects blood in urine.
16. Mention two causes of hematuria.
17. What is hemoglobinuria?
18. Name the tests used to detect bile salts and bile pigments in urine.
19. What are the conditions in which bile salts and bile pigments are present in urine?
20. What is jaundice? What are the different types?
21. What are the urinary findings of hemolytic and obstructive jaundice?
22. What is the role of bile salts in the body?
23. How are bile pigments formed in the body?

Identification of Unknown Abnormal Constituents in Urine

Urine sample containing abnormal constituents

→ **Benedict's test**
Coloured precipitate (+ve test)
Indicates the presence of glucose in urine (**Glycosuria**)

→ i) **Rothera's test**
Purple-coloured ring (+ve test)

ii) **Gerhardt's test**
Port wine colour (+ve test)
Indicates the presence of **ketone bodies** in urine. (**Ketonuria**)

→ i) **Heat and acetic acid test**
Coagulum is formed (+ve test)
Indicates the presence of albumins in urine (**Albuminuria or Proteinuria**)

ii) **Sulphosalicylic acid test**
White precipitate (+ve test)
Indicates the presence of **albumins** in urine

→ **Benzidine test**
A blue or green colour is developed. Indicates the presence of **blood** in urine (**Hematuria**)

→ i) **Hay's test**
Sulphur powder sinks to the bottom. Indicates the presence of **bile salts**.

ii) **Pettenkoffer's test**
Purple ring is formed (+ve test)
Indicates the presence of **bile salts** in urine.

→ i) **Fouchet's test**
Green or blue colour (+ve test)
Indicates the presence of **bile pigments** in urine.

→ ii) **Gmelin's test**
Green, blue or violet ring (+ve test)
Indicates the presence of **bilirubin** in urine.

5 Clinical Biochemistry Laboratory: Instrumentation and Techniques

BI 11.1; 11.6; 11.16; 11.18; 11.19

The qualitative and quantitative analyses of substances of biochemical importance in body fluids play a vital role in diagnosis, treatment and clinical monitoring.

The various body fluids subjected to chemical analyses include blood, plasma, serum, urine, cerebrospinal fluid (CSF), synovial fluid, amniotic fluid, pleural, pericardial and ascitic fluids. Other samples include solid tissues, calculi and feces.

The basic biochemical techniques employed in chemical analyses include pipetting, weighing micro quantities, reagent preparation and equipment management.

Glassware and Instruments Commonly Used in the Laboratory

Today, many items are made of plastic.

1. **Test tubes** are the most widely used glassware in the laboratory. Conical flasks are used for heating liquids. Round and flat-bottomed flasks are used for preparing solutions.
2. **Volumetric flasks** are graduated for taking the correct volume of liquids. They are used for preparing standard solutions. Beakers are available in different sizes.
3. **Graduated cylinders** are used for preparing reagents above the volume of 25 mL.
4. Different types of **pipettes** are used to deliver accurate quantities of liquids (Fig. 5.1).
   (a) **Graduated pipettes**: They are available in various sizes. They are of two types.
      (i) **Blow-out pipette**: A blow-out pipette is not completely graduated. For example, in a 10 mL blow-out pipette, graduation is present from 0 mL to 9.0 or 9.5 mL. The last portion of the fluid in the pipette (in the non-graduated portion) should also be delivered.
      (ii) **Non-blow-out pipette**: Unlike blow-out pipettes, non-blow-out pipettes are completely graduated. For example, a 10 mL pipette has markings from 0 to 10 mL. The fluid at the tip beyond the 10 mL marking should not be delivered.
   (b) **Volumetric pipettes**: These pipettes have a large bulb in their stem. The bulb increases the volume of liquid that the pipette can contain. These pipettes present as two types: blow type and non-blow type.
   (c) **Micropipettes and microsyringes** are used to transfer microquantities of samples (100 µL, 50 µL, 10 µL, 1 µL, etc., where 1 µL = 0.001 mL).
5. Various types of bottles like specimen bottles, reagent bottles and drop bottles are available for various purposes.

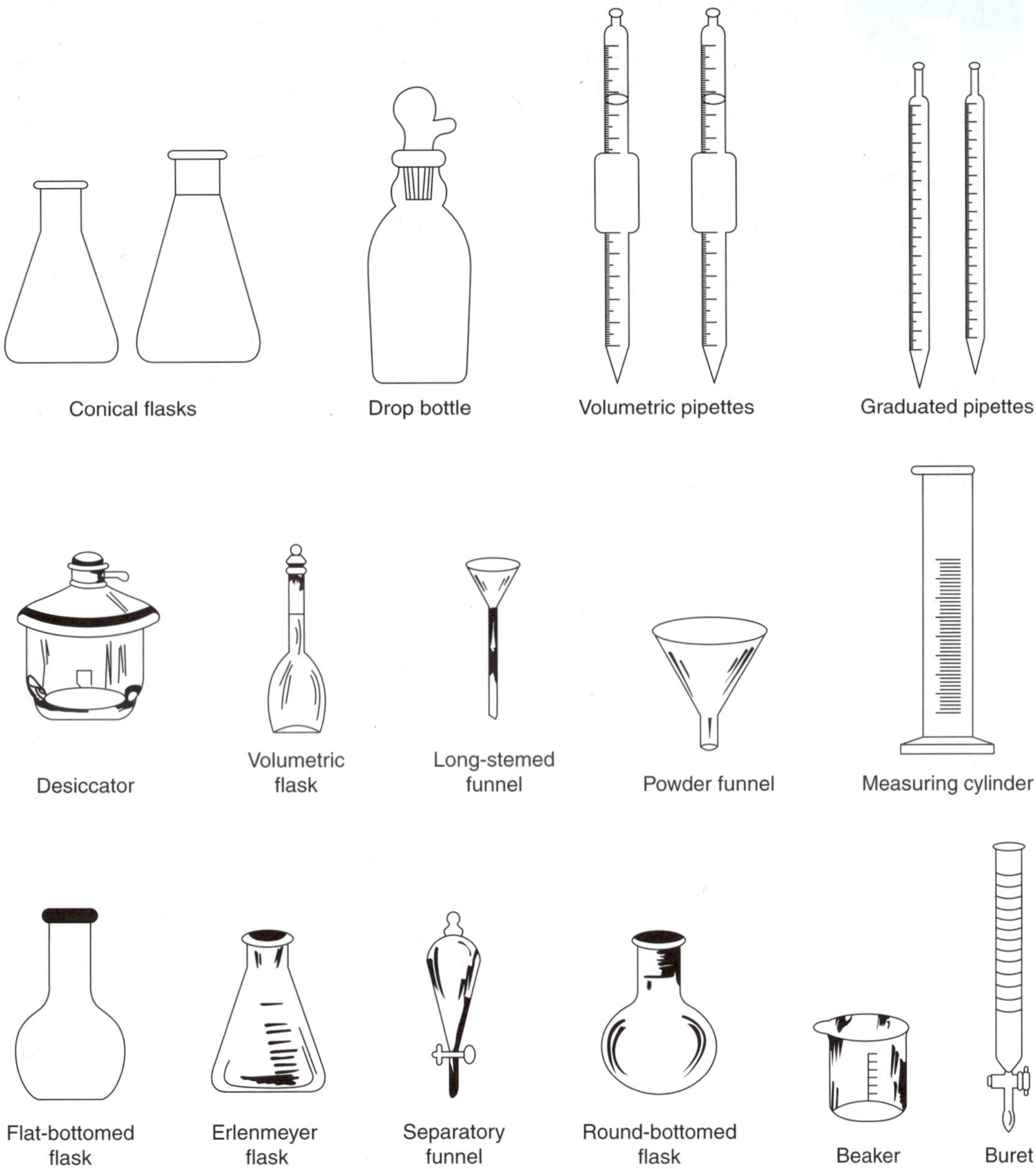

Fig. 5.1 *Glassware used in a biochemistry laboratory*

6. All glassware should always be clean and dry. Wash them thoroughly under running tap water after soaking them overnight in cleaning solutions like potassium dichromate. Dry them in the hot air oven at the appropriate temperature.

Weighing

Different balances are used in a biochemistry laboratory to weigh the chemicals used in the preparation of reagents. The ordinary scientific balance is used to weigh the chemicals which weigh more than 500 mg. The analytical balance is used for weighing chemicals which weigh less than 500 mg. During weighing of any chemical, the weight of paper or container in which they are kept for weighing should also be taken into account. Today, electronic analytical balances are available which are very accurate.

Centrifuge

Centrifuge is the instrument used to sediment particles such as cells which occur in a suspended state in fluid. Hand centrifuge, motor-driven centrifuge and micro hematocrit centrifuge are some of the types commonly used.

Hot air oven

It is used for drying and sterilising glassware. It is electrically heated and thermostatically controlled.

Incubator

This also uses electricity and regulates temperature thermostatically. It is used for various investigations where incubation is required.

Water bath

It is electrically heated and provides a range of temperatures from room temperature to 100°C.

Desiccator

Desiccators are used to dry or keep dry solid or liquid materials.

Indicators and pH scale

Certain acids or bases of weak strength are used as indicators in acid–base titration. They change colour depending on the hydrogen ion concentration in the solution. The change in colour is due to reversible ionisation of the indicator molecules. Ionisation of different indicators takes place at different pH ranges. For example, phenolphthalein is a colourless molecule; when it ionises it turns pink. This occurs at a pH between 8.0 and 9.8. So below pH 8.0, it is colourless. Above 8.0, up to 9.8, ionisation proceeds and different shades of pink are obtained. Beyond pH 9.8, no further change in the pink colour already obtained is observed.

pH is the negative logarithm of hydrogen ion concentration. The pH scale ranges 0–14. pH 7.0 is considered as neutral pH, below 7.0 is acidic and above 7.0 is alkaline.

Table 5.1 *Common indicators, their colour change and pH range*

Indicator	Colour change	pH range
Litmus	Red to yellow	5.5–8.0
Bromocresol green	Yellow to green	3.8–5.4
Phenolphthalein	Colourless to red	8.0–9.8
Methyl orange	Red to orange	3.0–4.4

Buffers

A buffer is a solution that can resist change in its pH upon the addition of a small amount of strong acid or base to it. They are used in certain experiments carried out at a particular pH. A buffer system is composed of a weak acid and and its conjugate base. For example, acetate buffer is prepared by adding sodium acetate to acetic acid. Bicarbonate buffer, phosphate buffer and citrate buffer are the commonly used buffer mixtures (see also Chapter 24).

Solutions

Different types of solutions are prepared in the biochemistry laboratory for various purposes. Their strengths can be expressed in different ways.

Saturated solution

A saturated solution can be prepared by continuously dissolving the solute, little by little, into the solvent until some amount of solute is left undissolved and remains as visible residue.

Percent solution

It is prepared by dissolving 1 g of solute in 100 mL of solvent. Percent solution can be expressed in three ways.

Weight per unit weight [W/W]

In this, the weight of the solute and solvent add up to 100, regardless of the final volume produced (the final solution need not have a volume of 100 mL). A 30% solution would have 30 g of solute dissolved in 70 g of solvent.

Weight per unit volume [W/V]

This is the commonly used method for preparing most laboratory solutions. The solute is usually a solid material; a weight of the solute is dissolved in a final volume of 100 mL. (dissolve the weight of the solute in 100 mL final volume and not in 100 mL of the solvent).

How to prepare a W/V solution

1. Take the weighed solute in a volumetric flask and add the solvent (ensure that the volume of solvent taken is less than the final volume).
2. Dissolve the solvent completely and make up the final volume by adding more of the solvent.

Volume per unit volume [V/V]

This is the method of choice when the solute is a liquid. A volume of the liquid solute is brought up to 100 mL of the final volume with solvent.

Molar solution

A molar solution is a solution in which 1 mole (gram-molecular weight) of the solute is dissolved in one litre of the solution. If 1 g of molecular weight is dissolved in a litre of a solution, it is termed '1 Molar' (1 M) solution. For example, the molecular weight of NaCl is 58.5. So **1 molar solution** of salt contains 58.5 g of NaCl dissolved in one litre of water.

Molal solution

It is prepared by dissolving 1 g molecular weight of a substance (1 mole) in 1 kg of solvent. For example, 40 g of NaOH is dissolved in 1000 g of water to make **1 molal** solution.

Normal solution

A normal solution contains a gram equivalent weight of solute dissolved in one litre of solution (**'1 N' solution**). For example, if the equivalent weight of a substance is 50, then 50 g is the gram equivalent weight, and when it is dissolved in water and made up to 1 L, it is a 1 N solution. Likewise, 100 g of that substance in 1 L is 2 N, 25 g of it in 1 L is a 0.5 N solution.

Sample Collection and Handling

Blood is the body fluid that is most commonly used for analytical procedures. Blood is obtained by skin puncture, venipuncture and sometimes by arterial puncture.

Blood will clot within a few minutes after it is drawn out unless an anticoagulant is added to stop the clotting. For hematology studies, unclotted **whole blood** is used. If blood is mixed with an **anticoagulant** and centrifuged, the cell components (such as RBCs and WBCs) are precipitated, leaving the supernatant **plasma.** Plasma is required in coagulation studies. **Serum,** the fluid portion of clotted blood, is used in a number of investigations.

What is the difference between serum and plasma?

If blood is withdrawn and allowed to clot without adding any anticoagulant to it, then after 2 hours or so the liquid portion separates from the clot. This is known as **serum**. In other words, plasma minus fibrinogen is serum. Serum lacks clotting factors like fibrinogen and prothrombin, as they are used in coagulation.

Centrifugation hastens the process of separation of serum. Though serum from coagulated blood is the specimen of choice for many investigations, plasma is a better sample in case of emergencies as the preparation of serum requires 15–30 minutes for clotting before centrifugation. Also, the yield of plasma from a given volume of whole blood is always greater than the yield from serum.

Sample collection

Syringes of different volumes are commercially available (for example, 2 mL, 5 mL, 10 mL and 20 mL). Select a disposable syringe of appropriate size depending on the amount of blood needed. If glass syringes are used, they should always be properly sterilised and completely dried. Sterilised sharp needles of bore size 18–20 gauge for adults and 23 gauge for children are recommended. The bevel length for adults should be medium (20 mm) and for children short (15 mm).

Venipuncture

The volume of blood obtained by venipuncture is sufficient to perform a number of tests. Whenever a venipuncture needs to be performed, a tourniquet must be used. The median cubital vein is the most frequently used vessel in venipunctures. Most biochemical tests are done on serum samples. Serum separated from whole blood should be collected in a clean, dry, sterile test tube. While drawing the blood, the tourniquet should not be held tightly for more than a minute.

Arterial puncture

Arterial blood is collected for blood gas analysis (arterial blood gases or ABG). Blood gas measurement is done to assess oxygenation problems in patients suffering from pulmonary embolism and pneumonia, and patients on ventilation to monitor extremes of oxygenation that produce either hypoxia with respiratory acidosis or oxygen toxicity. Patients suffering from serious cardiovascular disease and those undergoing cardiopulmonary surgery are monitored for hypoxemia.

Radial, brachial and femoral arteries are the usual sites for arterial puncture.

Skin puncture

Capillary blood is obtained by skin puncture. Skin puncture is preferred only when small quantities of blood are required to make a blood smear. Skin puncture is usually done by a finger prick.

Anticoagulants

Serum from coagulated blood is the specimen of choice for many tests. When whole blood or plasma is required for testing, an anticoagulant must be added to the specimen during collection.

Heparin is the most widely used anticoagulant. It acts by preventing the formation of thrombin from prothrombin and thus the formation of fibrin from fibrinogen.

Ethylene diamine tetraacetate (EDTA) is a chelating agent that prevents coagulation by binding to calcium. It is used in concentrations of 1–2 mg/mL of blood. It is used especially for hematological examinations, as it preserves the cellular components of blood.

Sodium, potassium and **ammonium oxalates** inhibit blood clotting by forming insoluble complexes with calcium ions. **Sodium citrate** is a widely used anticoagulant.

Sodium fluoride is used as a preservative of blood as well as an anticoagulant, usually in blood glucose estimations. It inhibits 'enolase', a glycolytic enzyme, and blocks glycolysis. Without the addition of an antiglycolytic agent, the blood glucose concentration decreases by about 10 mg/dL per hour at 25°C, and false low values may be obtained. Fluoride also is a potent inhibitor of the enzyme urease. This is why **sodium iodoacetate** is used instead of sodium fluoride as an anticoagulant when estimations of glucose and urea are performed on a single specimen.

Collection of CSF

Cerebrospinal fluid is drawn by lumbar puncture. Though up to 20 mL can be safely removed, this quantity is usually not required. CSF is examined in suspected cases of meningitis and cerebrovascular accidents.

Pleural, peritoneal (ascitic) and pericardial fluids

The pleural, pericardial and peritoneal cavities normally contain a small amount of serous fluid that lubricates the membrane surfaces. During infections, the fluids accumulate in large quantities. This can be removed by paracentesis to determine whether it is an effusion or an exudate. This can be done by protein enzyme analysis.

Synovial fluid

It is drawn from joint spaces. Synovial fluid is analysed to determine the type of arthritis and to differentiate non-inflammatory effusions from inflammatory fluids.

Amniotic fluid

Amniotic fluid drawn by amniocentesis is used to assess fetal maturity and to diagnose congenital disorders and intrauterine infection.

Examination of solid tissues

Examination of solid tissues is usually performed on malignant tissues. The tissue should be frozen as soon as possible in liquid nitrogen or in a mixture of ice and alcohol and later analysed.

Colorimetry

Estimation of biochemically important substances in body fluids (blood, CSF, urine, etc.) has tremendous clinical significance. It aids in arriving at a proper diagnosis, prognosis, follow up and monitoring of the treatment of various diseases. As some of these substances occur at extremely low concentrations, measuring them accurately is rather difficult. To determine the exact concentrations of these substances, a wide range of instruments is used in the clinical laboratory.

What is colorimetry?

Colorimetry is the technique of measuring the concentrations of substances based on light absorption by molecules in a solution. Various analytical instruments based on the principles of colorimetry are commonly used to measure substances like glucose, urea and creatinine in body fluids after converting them into coloured compounds. Concentrations of substances in various samples are determined by measuring light absorption at different wavelengths.

- Light is a form of energy. It can be both a particle and a wave. When light is considered as a wave, the distance between two successive crests is termed its wavelength, designated by λ (lambda) and expressed either in Å (angstroms) or in nm (nanometers).
- When ordinary light is passed through a prism, it splits into seven different colours by dispersion. The visible range of this light (VIBGYOR) will cover a wavelength from 400 nm to 700 nm.
- The colour of light is a function of its wavelength.
- When all types of radiations are arranged in the order of increasing wavelength, it is known as **electromagnetic spectrum** (Fig. 5.2).

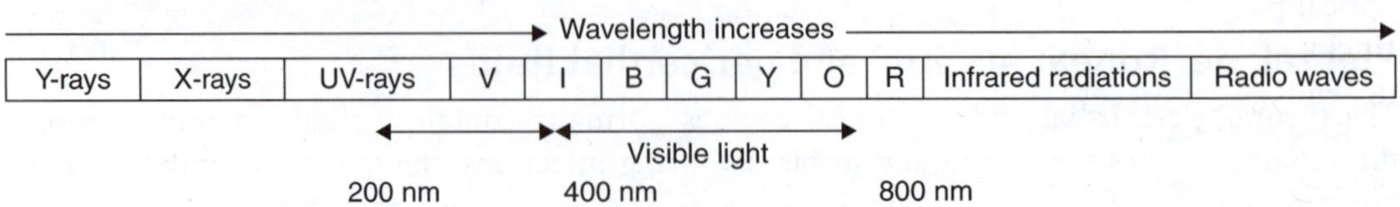

Fig. 5.2 *Electromagnetic spectrum*

Table 5.2 *Visible spectrum*

Wavelength (nm)	Colour of filter (colour of light absorbed)	Colour of solution (colour of reflected light)
400–435	Violet	Green-yellow
435–500	Blue	Yellow
500–570	Green	Red
570–600	Yellow	Blue
600–630	Orange	Blue-green
630–700	Red	Green

Principles of colorimetry

- Coloured solutions have the property of absorbing certain wavelengths of light (monochromatic light) and reflecting (transmitting) certain others. For example, a solution of hemoglobin absorbs blue-green light and reflects red and thus the solution appears red in colour.
- In colorimetry, filters are selected so that absorption is maximum at the chosen wavelength.
- In colorimetry, only coloured compounds or compounds capable of forming coloured derivatives by reacting with test reagents can be analysed. In these estimations, the colour is directly proportional to the concentration of the substance in the sample.
- In spectrophotometry, both coloured and colourless substances can be measured through ultraviolet spectral analysis.
- When light of a definite wavelength is passed through a coloured solution (incident light), some of it is absorbed, and therefore, the light that is emitted (transmitted light) is diminished. The extent of light absorption is determined by the following laws.

Beer–Lambert Law

Beer's law

This states that the absorption of light depends on the number of absorbing molecules through which the light passes. **The amount of transmitted light diminishes exponentially with an increase in the concentration of absorbing material.** Thus, the amount of light absorbed depends on the concentration of the substance in a coloured solution.

Lambert's law

According to Lambert's law, absorption of light is directly proportional to the thickness of the optical path. **The amount of transmitted light decreases exponentially with increase in the thickness of the medium.** In other words, the amount of light absorbed by a coloured solution depends on the length of the column or the depth of the solution through which light passes.

By combining these two laws (Beer and Lambert laws), the following equation can be obtained.

$$I_E/I_O = e^{-Kct}$$

where

I_E = intensity of the transmitted (emerging) light
I_O = intensity of the incident light
e = the base of natural logarithm (= 2.718)
K = constant
c = the concentration of the coloured substance
t = the thickness of the layer of the solution through which light passes

Transmittance (T)

The transmittance, T, of the solution is defined as the ratio of the transmitted intensity, I_E, over the incident intensity, I_0, and takes values between 0 and 1. However, it is more commonly expressed as percentage transmittance.

Percentage transmittance (T%)

$$T\% = \left(\frac{I_E}{I_O}\right) \times 100$$

Absorbance (A)

It is the negative logarithm of transmittance to the base ten.

$$A = -\log_{10} T\%$$

Optical density (OD)

The term optical density (OD) is the same as absorbance, i.e., it is the logarithmic ratio of the incident light (I_O) to that of transmitted light (I_E).

$$OD = \log_{10}\left(\frac{I_O}{I_E}\right) = -\log_{10} T\%$$

When transmittance is 100%, the optical density is zero.

Since optical density is in logarithmic scale, very low or very high readings will not give satisfactory results. Hence it is advisable to opt for those methods of test procedures that give the deflection on the OD scale between 0.10 and 0.60 (sensitive range).

When the length of the column through which the light passes (optical path) is kept constant (that is, Lambert's law is inoperative) by using cuvettes of the same diameter for both the test and the standard, the intensity of the transmitted light depends only on the concentration of the substance (that is, only Beer's law is operative).

Why should 'blank', 'standard' and 'test' solutions be prepared when testing a sample?

Most estimations of biochemical substances in blood and other body fluids are done by colorimetric methods. To measure a substance accurately, it is necessary to prepare blank (b), test (t) and standard (s) solutions.

How to prepare the 'test' solution

1. Take a specified volume of the blood filtrate or other specimen and mix it with reagents as mentioned in the respective procedure to produce the coloured derivative of the compound to be measured.
2. Take its OD reading in a colorimeter.

How to prepare the 'standard' solution

1. Prepare a solution of the pure substance (for example, glucose and urea) whose concentration is known.
2. Mix it with the reagents mentioned in the test procedure to produce the coloured derivative.
3. Take its OD reading.

The value of 'test' is compared with that of 'standard' of known concentration.

It is ideal to prepare a standard graph with different known concentrations.

How to prepare the 'blank' solution

It is not just the substance (to be estimated) in the sample that gives rise to coloured compounds (chromogens), but the reagents used in the procedure also contribute to the final colour developed. As a result, a value higher than the real concentration may be obtained. To nullify this, a **blank** is run by distilled water of the same volume (in place of the blood sample or standard solution), similarly treated with the reagents in the procedure, and an OD reading is obtained.

Subtraction of the OD reading of the 'blank' from that of 'test' and 'standard' gives the real reading corresponding to the concentration of the substance. Use the obtained OD readings of test, standard and blank in the following formula to estimate the concentration of the desired substance.

Calculation

Concentration of the substance (mg/100 mL)

$$= \frac{\text{OD of test} - \text{OD of blank}}{\text{OD of standard} - \text{OD of blank}} \times \frac{\text{Concentration of standard}}{\text{Volume of test}} \times 100$$

or

$$= \frac{\text{T} - \text{B}}{\text{S} - \text{B}} \times \frac{\text{Concentration of standard}}{\text{Volume of test}} \times 100$$

$$= \frac{\text{T}}{\text{S}} \times \frac{\text{Concentration of standard}}{\text{Volume of test}} \times 100$$

Colorimeter

The photoelectric colorimeter is widely used in clinical laboratories to estimate the concentrations of biochemically important substances in blood and other body fluids.

The basic components of a colorimeter include (Fig. 5.3):

1. **Light source** in the form of a filament lamp.

Fig. 5.3 *Components of a colorimeter*

2. **Filter**, which selects the light in a narrow wavelength (monochromatic light, mono = single, chrome = colour). In other words, it absorbs light of all other wavelengths in the visible range and allows only monochromatic light to pass through.
3. **Sample holder** known as '**cuvette**' made up of glass or quartz with a fixed diameter.
4. **Photo cell** such as photoelectric selenium cells, which are detectors commonly used in colorimeters.
5. **Display,** usually digital meter.

How it functions

The colorimeter works on the principle of the Beer–Lambert law. The instrument is operative in the visible range (400–750 nm) of the electromagnetic spectrum. The monochromatic light is allowed to fall on the coloured solution in the cuvette. The solution absorbs a part of light and the remaining emergent light falls on the detector photo cells. The electrical impulse which is generated as a result is measured and displayed appropriately.

How to operate a colorimeter

1. Choose the glass filter recommended in the procedure and place it in the filter slot. The colour of the filter should be complementary to the colour of the solution (see Table 5.2).
2. Take a cuvette and fill it to three-fourths of its volume with distilled water and insert it in the cuvette slot.
3. Switch on the colorimeter and allow it to warm up for 4–5 minutes.
4. Adjust to zero optical density.
5. Take the 'blank' solution in another tube and place it in the cuvette slot. Read the optical density as a reading of blank (B).
6. Take the 'test' solution in the cuvette, and, as with blank, read the OD for it as 'T'.
7. Lastly, take 'standard' solution in the cuvette and record the OD for it as 'S'. For accurate results, the OD values should be in the range 0.10–0.60.
8. Use the obtained OD readings in the following formula and calculate the concentration of the substance in the 'test' sample.

   Concentration of the substance (mg/100 mL)

$$= \frac{\text{OD of test} - \text{OD of blank}}{\text{OD of standard} - \text{OD of blank}} \times \frac{\text{Concentration of standard}}{\text{Volume of test}} \times 100$$

What is the difference between end point analysis and kinetic analysis?

While performing routine tests in the laboratory, the sample is mixed with the reagents and incubated at 37°C for a specific time, for example, 15 minutes, allowing the colour to develop optimally. After the incubation period is over, the OD reading is taken and the concentration of the substance is estimated. This is known as **end point analysis**. If, however, the sample and reagents are mixed and incubated for a short while, the OD readings are taken at 2 and 3 minutes exactly and the concentration is calculated by taking the difference between these two values, it is known as **kinetic analysis**. Because the readings are taken before completion of the incubation period, the colour is not developed properly. However, this type of analysis is used increasingly because it is quicker, especially in autoanalysers.

Spectrophotometer

A spectrophotometer is similar in its basic components to a photoelectric colorimeter (Fig. 5.4) and functions similarly. But it is more sensitive (*about one thousand times!*) than the colorimeter. Also, it can operate both in the visible region (400–750 nm) and the ultraviolet region (200–400 nm) of the electromagnetic spectrum. The individual components of a spectrophotometer, although similar to that of a colorimeter, are more sophisticated and it costs much more than an ordinary colorimeter. Instead of the usual glass cuvettes, quartz cuvettes are used in the UV light range. A hydrogen discharge lamp replaces the tungsten filament lamp for the UV region. Concentrations up to micrograms can be estimated accurately in a spectrophotometer.

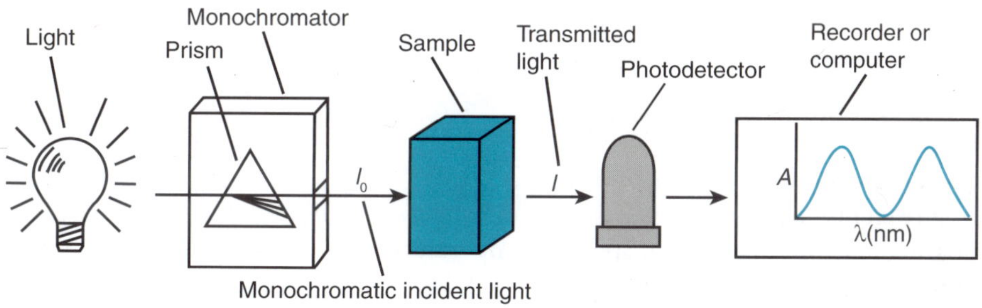

Fig. 5.4 *Schematic of a spectrophotometer*

Types of spectrophotometer

1. Single beam
2. Double beam

Uses of spectrophotometer

- Can be used for routine clinical chemistry analysis for the identification of compounds in mixtures
- Concentrations up to micrograms can be estimated accurately in a spectrophotometer.

Autoanalyser

This is an ultra modern, highly sophisticated equipment in which a number of samples can be analysed within a short span. Most laboratories use semi-automated analysers in which some steps like pipetting the sample, reagent mixing and incubation are done manually. Other steps involving aspiration of the coloured solution, measurement, calculation and display/printing are automated.

Fully automated systems, though expensive, are very convenient to use and save time. Batch analysers perform for one constituent in several samples, whereas discrete selective analysers, which are highly sophisticated systems, can analyse a number of parameters in a single sample simultaneously. One can choose any number of desired parameters on a sample.

Features

- Autoanalysers make use of automation to analyse specimens. They enable the process of analysis of large numbers of specimens without a comparable increase in laboratory staff.
- The international union of pure and applied chemistry (IUPAC) describes automation as the replacement of human manipulative efforts and facilities in the performance of a given process by mechanical and instrumental devices that are regulated by feedback of information so that an apparatus is self-monitoring (or) self- adjusting.
- The automation in an autoanalyser is a self-regulating process, wherein the specimen is accurately pipetted by a mechanical probe, mixed with a particular volume of reagent and results displayed in digital format as well as printed by a printer.

Automation

The simple meaning of automation here is the mechanisation of various duties performed by laboratory personnel consisting of:

1. Pipetting of reagents
2. Pipetting of specimen
3. Standards, calibrators, quality control (QC), sera, etc.
4. Mixing reagents and specimens (reagent and standard, calibrator)
5. Incubating reaction mixtures at specific temperatures
6. Reading intensities of end products
7. Calculating test results
8. Printing test reports
9. Preparing QC data and charts

Advantages of autoanalysers

1. Large number of samples can be tested in a short time.
2. Variety of tests can be performed by using various methods such as end point and rate of reaction analysis.
3. The concentration of hormones, drugs, tumour markers and various types of antibodies and proteins are determined using techniques such as ELISA, chemiluminescence, turbidimetry, nephelometry, etc.
4. The methods performed with automation are accurate, precise, sensitive and specific.
5. Cost-effective, because the amount of reagent and specimens are as low as 100 μL and 50 μL respectively.
6. Automation allows laboratories to process much larger workload without a comparable increase in personnel.
7. Internal and external QC programmes can be implemented efficiently and effectively using autoanalysers.
8. In a fully-automated analyser, the laboratory operator does not come in contact with specimens and reagents. Hence, working with these analysers is very safe.

Broadly, autoanalysers can be categorised into (i) continuous-flow analysers or (ii) discrete analysers.

Continuous-flow analyser: In these, samples and reagents pass sequentially through the same analytical pathway followed by mixing in a common pathway. In a single-channel continuous flow analyzer, at a time only one type of test is performed and it is possible to test a large number of specimens for a particular test accurately in a short time. In a multi-channel continuous flow analyser, 6–12 individual tests of nearly 60 specimens are conducted simultaneously.

Discrete analyser: These can run multiple tests on one sample at a time or one test on multiple samples at a time.

The basic features these analysers are:

1. The laboratory technician/operator plays a role in its functioning.
2. Reagents and specimens are pipetted in a cuvette, mixed and the end products are recorded photometrically.
3. All end-point rate of reactions can be performed in monochromatic (one filter/wavelength) (or) bichromatic (2 filters/ wavelengths) modes.
4. More than 100 open test programs can be selected.
5. They are economical since the reagent quantity required is only 200–500 μL/test.
6. These analysers offer "patient use" report.
7. A large number of test results can be stored.
8. Substances analysed include certain metabolites, electrolytes, proteins and drugs. Hormone assays and routine chemistry tests can also be performed.
9. Nonlinear curve plotting is possible for hormonal assays.

Discrete analysers are of two types:

1. Semi-automated analysers
2. Fully-automated analysers

Semi-automated discrete analysers

Semi-automated analysers are so called because the initial stages of the specimen analysis such as pipetting of the reagent and specimen, mixing and incubating the reaction mixture are performed by laboratory technicians. These analyses allow selection of routine and special biochemical tests programs directly through keyboard. The following functions are performed by the semi-auto analysers:

- Reading end point (linear and nonlinear) and rate of reaction (visible and UV-kinetic) in monochromatic and bichromatic modes.
- Displaying test results, printing and memorizing these results and graphs of all linear and non-linear reactions.

All standard operation procedures (SOPs) are stored in memory. The software is user friendly and the system comes with backlit LCD display and battery backed-up power system for functioning when there is no power. Other functions facilitated include statistical analysis to determine mean values, SD, CV for control samples and their plots, immunoassays, and coagulation and drug assays.

Fully-automated discrete analysers

These analysers perform all the functions of a semi-automated analyser, and in addition, they perform the following functions:

- Automatic dispensing of reagents (probe)
- Automatic dispensing of samples (sampler) by means of a specimen probe.
- Automatic mixing of reaction mixtures
- Incubating reaction mixtures

These analysers are equipped with the following additional features:

- A sampler to accommodate various samples, standards, calibrators and quality control serum
- Programmable sample probes
- Reagent probes for the pipetting of reagents
- Reagent carousel—contains various reagent containers
- Syringes, water and waste bottles for easy accessibility and maintenance
- Reusable reaction cells
- Programmable wash cycles between samples and tests for minimum carryover
- Cuvette rinsing station – the advanced versions of the fully automated analysers are to clean each cuvette with detergent and deionized water before use. This ensures excellent precision and reliability of results.

The components/steps in fully automated systems are given below.

1. **Specimen identification**: A bar-coded label is placed on the specimen container and read subsequently by one or more barcode readers placed at key positions in the analytical pathway. Using this method, the error rate in specimen identification can be reduced and the identity of a specimen can be verified with greater accuracy.
2. **Specimen preparation, handling and delivery**: Adequate quantity of blood is collected in vacutainer tubes after centrifugation and the tubes are placed in the sample carousel. The specimen probe is equipped with level sensor; the required amount of serum is pipetted automatically. The use of level sensors restricts the penetration of sample probes into specimens, provides smoother motion control and eliminates spatter. Disposable specimen cups (made up of glass and polyvinyl) and the design of the device for holding specimens is specific for a specific instrument. It may be in the form of a tray, rack or chain.
3. **Reagent handling, storage, identification and delivery**: Liquid reagents stored in plastic or glass containers are used in most automated systems. Single reagent (mono-step method) is used; two or more reagents are required for some tests. Labels on reagent containers include information for reagent identification, volume of the contents, expiry date and lot number. Most of the reagents use barcode for capturing this information. Liquid reagents are taken up and delivered to the mixing and reaction cups or cuvettes by pumps via tubes or by positive displacement syringe devices. Syringe devices are used in many automated systems for both reagent and sample delivery. The reagent probes are washed and flushed if more than one reagent is used to prevent reagent carryover.
4. **Mixing, incubating and reading**: The mixing of specimen and reagents takes place in thermo-cuvettes by techniques that include forceful dispensing, magnetic stirring by a vibrating rod, rotating paddle and the use of ultrasonic energy. For incubation, the analysis

compartment is placed in a large air bath or water bath heated at 30°C to 37°C. The cuvettes are transparent in the spectral range and are measured automatically.

5. **Measurement devices**: For the measurement of the end result, the following techniques are employed:
   - Photometry: Spectrophotometry (and reflectance photometry)
   - Turbidimetry: Nephelometry
   - Ion-selective electrodes and other electro chemical techniques
6. **Signal processing, handing of data and control of microprocessors**: The digital computers in automated systems convert analogue signals from detectors to digital form by analogue-to-digital converters.

The microprocessors enable the following:

- All electrochemical operations to be performed uniformly, repeatedly and in correct sequence. These include transfer of solutions, placement of proper filters for bichromatic (or) multi-chromatic reading, regulation and change in speed of rotation of carousels.
- Acquire, process, assess and store operational data from the analysers.
- Important communication interactions related to trouble-shooting
- Output data tested against present criteria is flagged and displayed for the operator's evaluation and judgement. Such information specifies if
  - linearity of a reaction has been exceeded
  - a reaction is nonlinear
  - substrate exhaustion has occurred
  - absorbance of end product is very low/very high
  - baseline drift is excessive
- The microprocessors can take command from the operator through the keyboard or a touch-sensitive screen for
  - temperature set point
  - calibrator concentration
  - format of printed output and quality control charts

Flame Photometer

It is an instrument **for estimating the concentration of electrolytes** such as sodium and potassium. When the solution is sprayed over a flame, it dissociates to generate neutral atoms. Some of these get excited and move to a higher plane of energy. When excited atoms fall back to the ground state, they emit light of a characteristic wavelength specific to that particular element, which can be measured.

More recently, ion selective electrodes, expensive but more sensitive, are used in estimating electrolytes such as sodium, potassium and lithium.

Fluorimeter

The fluorimeter is based on the principle of **fluorimetry**, in which the substances to be measured are exposed to light of a particular wavelength. Upon exposure, these substances get excited, and while returning to the ground state, emit light in the form of fluorescence proportional to the concentration of the substance.

Dry Chemistry Systems

In these systems, all the reagents necessary for a reaction are embedded in a plastic matrix in the dry state and are used in emergency practice. Today, dry chemistry systems like glucometers are widely used by diabetic patients for glucose estimation at home.

ABG Analysis

Arterial blood gases (ABGs) are measured in laboratory to assess **blood gas**, pH, electrolytes, and some metabolites in **whole blood specimens** to detect acid–base imbalance leading to disorders such as respiratory acidosis, respiratory alkalosis, metabolic acidosis and metabolic alkalosis. An ABG analyser can measure pH, partial pressure of carbon dioxide and oxygen (pO_2, pCO_2), and concentrations of many ions (sodium, potassium, chloride, bicarbonate) and metabolites (calcium, magnesium, glucose, lactate).

How do you check ABG?

As blood passes through lungs, oxygen moves into the blood while carbon dioxide moves out of the blood into the lungs. Blood for an **ABG test** is taken from an artery (radial artery in the wrist, femoral artery in the groin, brachial artery in the arm), where the oxygen and carbon dioxide levels can be measured before they enter body tissues.

How does an ABG machine work?

The arterial blood collected from the patient is introduced into the analyser. Modern blood gas analyzers are fully automated machines capable of measuring various forms of hemoglobin, serum electrolytes, and metabolites. The analyser aspirates the blood into a measuring chamber that has ion selective electrodes, sensitive only to the measurement of interest—pH, pCO_2, pO_2, HCO_3 level. After measurement, the blood is automatically expelled into a waste container and the sample path is cleaned, ready for the next sample. Results are displayed and printed.

Normal levels of ABG

According to the National Institute of Health, typical **normal values** are:

pH	:	7.35–7.45.
pO_2	:	75–100 mmHg
pCO_2	:	35–45 mmHg
HCO_3 range	:	22–28 mmol/L

Blood Gas Analysis

The determination of pO_2, pCO_2 and pH play an important role in the detection of acid–base imbalances and their monitoring and treatment. Arterial or venous blood specimens (obtained commonly from the vessels of the extremities) are used for this purpose. Since there is an inherent risk in taking samples frequently for blood gas measurements, particularly in case of arterial punctures and sample collections from premature infants, several noninvasive methods have been developed that have proven effective in the continuous transcutaneous monitoring of pO_2 and pCO_2.

Differences in measured blood gas values between the arterial and the venous are more marked in case of pO_2. pO_2 is about 60 mm Hg lower in venous blood after oxygen is released from the tissue capillaries, whereas pCO_2 is about 6 mm Hg higher in the venous blood. pH is 0.02 to 0.05 pH units lower in venous blood.

Arterial Blood Gases (ABG) in Acid–Base Disorders

The measurement of blood gases is of immense value in assessing acid–base status. Arterial blood is used to measure the acid–base parameters such as pH, pO_2 and pCO_2. The blood is collected, usually from the radial or the femoral artery (air-free), and is analysed quickly within 30 minutes. The interpretation of the results will reveal whether any acidosis or alkalosis is due to respiratory malfunction or metabolic abnormality (Table 5.3).

Table 5.3 *Arterial blood gas (ABG) analysis reference ranges**

Parameter	Reference range	
	SI units	Non-SI units
H^+ ion	37–45 nmol/L	pH 7.35–7.45
HCO_3^-	22–28 mmol/L	22–28 mEq/L
Base excess	–2 to +2 mmol/L	–2 to +2 mEq/L
pCO_2	4.5–6.0 kPa	35–45 mm Hg
pO_2	12–15 kPa	90–113 mm Hg
Oxygen saturation	>97%	>97%

*Note that HCO_3^- concentration is derived from pH and pCO_2 values.

Immunodiffusion

Immunodiffusion is a diagnostic test involving the diffusion of antigen and antibody through a substance such as agar. Generally, soft gel agar (2%) or agarose (2%) are used for the detection of antibodies or antigen. Agarose is the more preferred medium over agar.

The rate of diffusion mainly depends on

- Size of particles
- Temperature
- Viscosity
- Amount of hydration
- Interaction between matrix and reactants.

The commonly known types of tests are:

1. Single diffusion in one dimension (Oudin procedure)
2. Double diffusion in one dimension (Oakley_Fulthorpe procedure)

3. Single diffusion in two dimensions (radial immunodiffusion or Mancini method)
4. Double diffusion in two dimensions (Ouchterlony double immunodiffusion).

Principle

In aqueous soution, the soluble antibody (Ab) and the antigen (Ag) interact to form a lattice that develops into a insoluble visible precipitate. Soluble Ag could be toxins, toxoids, proteins, carbohydrates, glycoproteins, or lipoproteins. The formation of Ag–Ab lattice depends on the valency of both Ag and Ab.

Soluble Ag + soluble Ab = Precipitation.

Radial immunodiffusion immunoassay

- It is one of the gel-based methods for quantitative immunochemical study. With this technique, a concentration gradient is established for a single reactant, usually the antigen. The antibody is uniformly dispersed in the gel matrix.
- The antigen is allowed to passively diffuse from a well into the gel, and immune precipitation occurs until antibody excess exists. The antigen–antibody interaction is manifested by a defined ring of precipitation around the antigen well, the diameter of which will increase with increased antigen concentration. Calibrators are run at the same time as the sample, and a calibration curve of the ring diameter versus concentration is plotted (refer to Fig. 5.5). This curve is used to determine the concentration of the control and unknown samples.

In **Ouchterlony double diffusion,** the antigen and antibody diffuse through semisolid agar medium independently. The process is:

Wells are cut in the gel → Reactants are added in the wells → Incubate in a moist chamber for 12–48 hours → Preciptin lines will form where the moving front of the antigen meets the antibody

The **density of the line** reflects the quantum of immune complexes formed (Fig. 5.6).

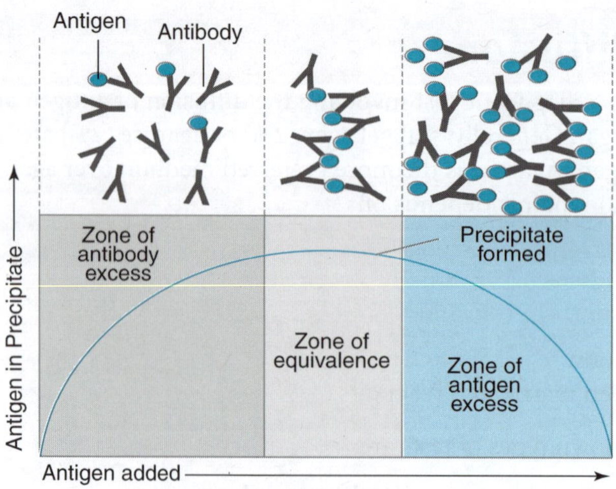

Fig. 5.5 *Calibration curve in radial immunodiffusion*

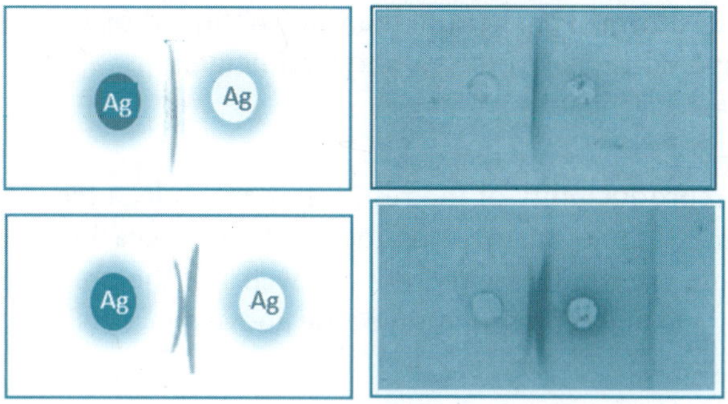

Fig. 5.6 *Ouchterlony double diffusion showing different quantum of immune complexes formed*

Applications of immunodiffusion

- To determine relative concentration of antibodies/antigens
- To compare antigens or determine the relative purity of an antigen preparation
- For disease diagnosis
- Serological surveys

Quality Control

Quality control (QC) is necessary to maintain the reliability and correctness of investigations performed in a laboratory. The quality control program are implemented in two ways: Internal and external.

Internal quality control

This refers to the set of procedures undertaken by the laboratory staff for the continuous monitoring of laboratory work to decide whether the results are reliable enough to be released. In this programme, the routine laboratory tests are assessed and evaluated on a daily basis. This can be done by preparing a reference standard sample and cross-checking the results of tests with it on a daily basis.

External quality control

In this programme, a central reference laboratory sends a serum sample of known concentration to the peripheral (small) laboratory. The sample is analysed at the peripheral laboratory and if the result is the same or close, the procedures followed in the peripheral laboratory are said to be reliable. Smaller laboratories tie up with reference laboratories for this type of quality control programme and it can be done at regular intervals, say once in two weeks.

Now let us discuss certain basic terms concerning quality control.

Accuracy

It is the measure of the closeness of the estimated value to the true value. For example, two technicians performing a test on a serum sample which is known to contain 100 mg/dL glucose obtain a result of 96 mg/dL and 120 mg/dL. The value obtained by the first technician is said to be more accurate compared to that of the second technician.

Precision

It is a measure of the reproducibility of the result. For example, a technician performs glucose estimation on the same sample on three or four different occasions and obtains 90 mg/dL, 88 mg/dL and 92 mg/dL; the results are reproduced well and the precision is said to be good.

Refer to Fig. 5.7 for a visual representation of the above concepts.

Specificity

Specificity means that just one substance should answer a particular test. For example, while estimating glucose by the glucose oxidase method, the method is specific to glucose and does not detect other reducing sugars (such as galactose) that react with other reagents.

Sensitivity

It denotes whether a method is capable of measuring substances occurring in extremely low concentrations or in dilute solutions. For example, the ELISA method is utilised to detect protein even if it occurs in a few micrograms. Usually, when sensitivity increases, specificity decreases.

$$\text{Sensitivity} = \frac{\text{True Positives}}{\text{True Positives} + \text{False Negatives}}$$

$$\text{Specificity} = \frac{\text{True Negatives}}{\text{True Negatives} + \text{False Positives}}$$

Laboratory quality control charts

Control charting is a form of internal quality control. The control chart for a given variable has control limits that define the acceptable range of the monitored variable. The laboratory

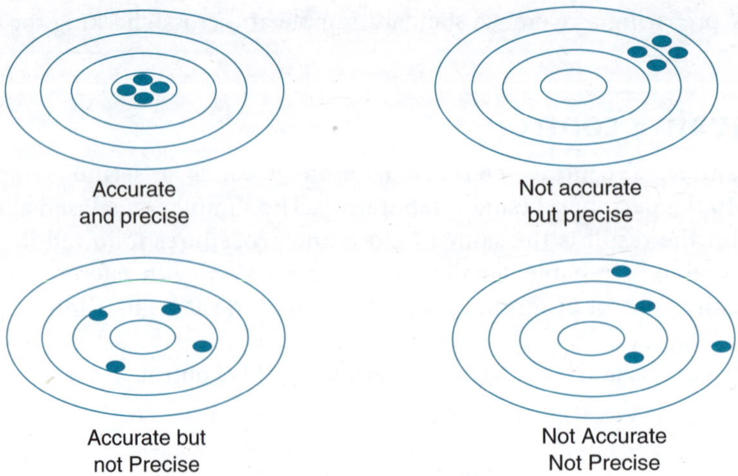

Fig. 5.7 *Diagramatic representation of precision and accuracy*

runs control samples together with the routine samples. The control values are plotted in the control chart. If the control value is outside the limits, remedial actions have to be taken to identify the sources of error and to remove such errors. Thus the control chart helps to detect accuracy problems in a laboratory's internal analytical process with the help of a visual display and thereby improve the quality of the results reported by the laboratory.

Levey–Jennings charts

These charts are similar to the individual control charts except that control values are set within the limits of ±3 SD. Control specimens are analysed during each run and the value plotted on the control chart. If the control values are within the control limits (i.e., + or –3 SD or 3 σ), patients' values can be reported. Even if a single control value is out of limit, reporting should be stopped, the problem resolved and the entire run of specimen and controls repeated.

Systematic error

Systemic error is evidenced by a change in the mean of the control values. It can result in shift or a trend. A trend indicates a gradual loss of reliability in the test system. Trends are usually subtle.

Random error

Technically, random error is any deviation away from an expected result. For QC results, any positive or negative deviation away from the calculated mean is defined as random error. Some examples are:

- Double pipetting of control sample
- Air bubbles in water supply
- Air bubbles in sample reagent or in pipetting system
- Insufficient volume of specimen
- Interfering substance present
- Instrument precision problem
- Wrong patient ID
- Reporting delay
- Transcription errors
- Report not legible

Westgard quality control rules

This uses the same procedure of quality control using two different controls with control limits of 2SD and 3SD values. When both the control values are within 2SD, the values are acceptable. If a single value exceeds plus or minus 2SD, it is to be taken as warning. If one value exceeds the plus or minus 3SD, random error is detected as per rejection rule. Two observations beyond plus or minus 2SD also is rejection rule sensitive to systemic error. If 4 values exceed the plus or minus 2–3 SD range, it indicates systemic error. This chart, therefore, helps to detect the type of error also. If the values are out of control, reporting should be stopped.

There are a series of control rules for interpreting control data to caution about fault in accuracy or precision of the measurement system. A convenient notation, such as 1_{2s} (indicating one measurement exceeds 2 SDs either above or below the mean of the reference range), 2_{2s} (indicating two measurement exceeds 2 SDs either above or below the mean of the reference range), etc, is adopted for stating the rules.

External quality assessment schemes (EQAS)

These refer to a system in which the performance of a laboratory is assessed and compared with a peer group of laboratories to indicate the accuracy of the laboratory's test results.

While internal QC maintains checks on the accuracy and precision of an analytical method, EQAS compare the performance of different laboratories. EQAS are necessary for maintaining long-term accuracy of analytical methods. The basic program involves the analysis of the same QC material by different laboratories and the results are dispatched to the central control laboratory for data analysis. A summary of reports is then sent to the participating laboratories by the controlling laboratory. For each analyte, the standard deviation index (SDI) is calculated using the student "t" test. When the difference is significant for a particular laboratory, that laboratory is alerted.

$$\text{SDI} = \frac{\text{Lab result} - \text{Group's mean}}{\text{Group's SD}}$$

Viva Questions

1. What are molar, molal and normal solutions?
2. What is the difference between serum and plasma?
3. What is colorimetry?
4. What is the Beer–Lambert law?
5. Why do we need to prepare a 'blank' when testing a sample?
6. What is the difference between end point analysis and kinetic analysis?
7. What is the quality control programme used in laboratories?
8. What are the uses of ABG analysis?
9. What is spectrometry?
10. What is automation? What are the different types of autoanalysers?
11. What is immunodiffusion?
12. What is quality control?What are the different types of QC?
13. Define accuracy, precision, specificity and sensitivity.
14. What are L-J charts?
15. What are Westgard QC rules?

6

BI 11.17, 11.21

Blood Glucose

Glucose is the body's chief source of energy. Estimation of glucose is commonly done in clinical laboratories to screen for diabetes, a condition in which the deficiency of insulin or resistance to its action causes high glucose levels in blood (hyperglycemia). The term blood sugar represents, besides glucose, other hexoses as well (in the post-absorptive state). But for all practical purposes, the term blood sugar is used synonymously with blood glucose.

Blood sugar estimated at any time of the day without any prior preparation is known as **random blood sugar (RBS)**. Sugar estimated in the morning before taking breakfast after an overnight fast of 10–14 hours is called **fasting blood sugar (FBS)**. The analysis carried out 2 hours after a meal is called **post lunch or postprandial** (Latin, after food) **blood sugar (PLBS or PPBS)**.

Sample Collection

The estimation of glucose concentration in plasma is preferable to that of whole blood because of its higher water content which contains more dissolved glucose. Plasma glucose concentration is 10–15% greater than that of whole blood because a given volume of RBCs contains less water than the same volume of plasma. The difference is of little consequence at normal concentrations; however, it is of importance when interpreting the results of glucose tolerance tests.

Erythrocytes continue to utilise glucose in vitro; as a result, unless the blood sample is analysed immediately, low values of glucose are obtained. Unless the plasma is separated from the blood cells within an hour of collection, it is essential that the whole blood sample be mixed with an inhibitor of glycolysis (such as sodium fluoride which inhibits the enzyme enolase) along with an anticoagulant (potassium oxalate).

Determination of Glucose in Body Fluids

Various procedures are available to measure the blood glucose level.

Chemical methods

Many of the relatively older procedures for measuring glucose are based on its property of acting as a reducing agent in hot alkaline solution. In hot alkaline solution, the aldehyde group of glucose reduces cupric ions to cuprous ions, which form cuprous oxide (Cu_2O). The cuprous oxide formed is made to react with phosphomolybdate to produce a coloured complex which is measured on a colorimeter.

As these methods utilise the reducing property of glucose, the presence of certain other reducing substances in blood such as ascorbic acid and uric acid may result in values up to 20–30 mg/dL higher than the real values.

The following are a few oxidation-reduction and condensation methods:

1. Folin–Wu method
2. Somogyi–Nelson method
3. King–Asatoor method
4. Ferricyanide method
5. Ortho toluidine method

All these methods are no longer in use and are of historical importance only.

Enzymatic methods

As these methods are based on enzymatic action, they are highly specific for glucose and do not measure any other substance, and hence the true value of blood glucose can be estimated. No heating is involved in enzymatic methods and hence can be used in autoanalysers.

The following are some of the enzymatic methods.

Hexokinase method

Glucose is converted to glucose-6-phosphate by hexokinase. The reaction is coupled to the action of glucose-6-phosphate dehydrogenase (G-6-PD) that uses NAD+/NADP+ as electron acceptor. Glucose-6-phosphate is oxidised to 6-phosphoglucono-lactone and NAD^+/$NADP^+$ is reduced to NADH/NADPH. The reduced coenzyme can be quantitated by measuring the absorbance at 340 nm. The amount of NADH/NADPH produced is directly proportional to the amount of glucose in the sample (G6PD derived from bacteria is used in the assay with NAD^+ as the cofactor; $NADP^+$ is the cofactor if G6PD derived from yeast is used).

Glucose oxidase method

This is the most commonly used method in most Indian laboratories (see the section "GOD-POD Method" that follows).

Semi-quantitative measurement of glucose in urine

Several glucose-specific paper test strips (dipsticks) are commercially available (clinistix, diastix, chemstrip, etc.). All use the glucose-specific enzyme glucose oxidase in a chromogenic assay.

The test end of the strip is moistened with freshly voided urine and examined after 10 seconds. Colour develops if glucose is present.

Glucose dehydrogenase method

Performing tests for glucose estimation routinely and repeatedly in patients both for initial diagnosis and for monitoring the treatment is painful and inconvenient. Attempts have been made to develop a painless method for estimating glucose concentrations. The following approaches may serve the purpose.

Implanted sensors: An electrochemical sensor that uses glucose oxidase implanted intravenously or subcutaneously is used for monitoring blood glucose levels.

Non-invasive glucose monitoring

1. **Infrared spectroscopy**: These devices measure either the absorption or the reflection of light from subcutaneous tissue. But this is not 100% specific for glucose.
2. **Iontophoresis**: In this method, an electrical current promotes movement of molecules across the skin.

Most of these non-invasive methods are in the experimental stage and are not likely to be used in routine patient care in the immediate future.

Self-analysis of glucose by the patient at home

Blood glucose concentrations measured using glucose-sensitive reagent strips by patients at home is becoming increasingly popular among diabetic patients.

Application of capillary blood by a finger prick to these strips causes a colour change proportional to the glucose concentration. After the blood has been washed off, the glucose concentration is read using a colour comparator or measured by reflectance photometry by a small electronic instrument **(glucometer)**. However, reagent strips should not be relied upon for the initial diagnosis of diabetes. Laboratory estimation is always preferable.

Estimation of Glucose by Glucose Oxidase (GOD-POD) Method

The standard glucose oxidase photometric method is highly specific for β-D-glucose. It is the method commonly used in most clinical laboratories and is simple to perform.

Principle

The glucose in the plasma is oxidised by glucose oxidase to gluconic acid and hydrogen peroxide.

$$\text{Glucose} + 2H_2O + O_2 \xrightarrow{\text{Glucose oxidase}} \text{Gluconic acid} + 2H_2O_2$$

The addition of enzyme peroxidase and a chromogenic oxygen acceptor, such as O-dianisidine, results in the formation of a coloured compound which is measured.

$$H_2O_2 + \text{Chromogen (colourless)} \xrightarrow{\text{Peroxidase}} H_2O + \text{Coloured product}$$

The H_2O_2 is split into H_2O and nascent oxygen by peroxidase. The oxygen oxidises the colourless chromogen to a coloured one. The intensity of the colour is directly proportional to the concentration of glucose.

Although the glucose oxidase (GOD) reaction is highly specific, the second step involving peroxidase (POD) is less specific. Various reducing agents in the blood like ascorbic acid, uric acid, bilirubin, hemoglobin, tetracycline and glutathione inhibit the reaction by competing with the chromogen for H_2O_2.

Procedure

1. Take three clean and dry test tubes. Label them **blank (B), standard (S)** and **test (T)**.
2. Add the solution to the test tubes, as given in the following table.

Reagents	Blank	Standard	Test
Distilled water	0.05 mL	–	–
Working standard	–	0.05 mL	–
Sample	–	–	0.05 mL
GoD-POD reagent	5.0 mL	5.0 mL	5.0 mL

Mix well and keep in an incubator for 15 minutes and take the OD (optical density) reading at 540 nm.

OD			

Calculation

Apply the values and readings obtained in the following formula.

$$\left.\begin{array}{c}\text{Concentration of glucose}\\ \text{in plasma (mg/dL)}\end{array}\right\} = \frac{\text{OD of test} - \text{OD of blank}}{\text{OD of standard} - \text{OD of blank}} \times \frac{\text{Concentration of standard}}{\text{Volume of test}} \times 100$$

The concentration of glucose standard is 100 (mg/dL); the working standard taken in the test is 0.05 mL, which contains 0.05 mg glucose.

$$= \frac{\text{OD of test}}{\text{OD of standard}} \times \frac{0.05\ \text{mg}}{0.05\ \text{mL}} \times 100\ \text{mg/dL}$$

$$\text{Concentration of glucose (mg/dL)} = \frac{\text{T}}{\text{S}} \times 100\ \text{mg/dL}$$

Estimation of Glucose by O-Toluidine Method

Principle

Aldohexoses (glucose) react with ortho toluidine in glacial acetic acid to form an N-glycosyl amine derivative, which is green in colour.

Reagents

1. Trichloroacetic acid 10%
2. O-toluidine reagent: Add about 90 mL of O-toluidine to 5.0 g of thiourea and dilute to 1 L with glacial acetic acid. Store in a brown bottle.

Procedure

1. Take 0.5 mL of the blood sample and add 3.0 mL of distilled water in a dry test tube.
2. Add 1.5 mL of 10% trichloroacetic acid. Mix well and wait for about 10 minutes and filter.
3. Take three clean and dry test tubes and mark them as **blank (B), standard (S)** and **test (T)**, and add the reagents as shown in the following table.

Reagents	Blank	Standard	Test
Filtrate	–	–	1 mL
Standard glucose (10 mg/dL)	–	1 mL	–
Water	1 mL	–	–
O-toluidine reagent	5 mL	5 mL	5 mL

4. Mix the contents of each test tube and keep in a boiling water bath for 10 minutes.

5. Cool and take the OD reading at 625 nm.
6. Calculate the result in the same way as shown in the GOD-POD method.

Interpretation

In normal healthy persons:

Fasting plasma glucose is 70–110 mg/dL.
Postprandial glucose is less than 140 mg/dL.

When blood glucose concentration is within the normal range, it is referred to as normoglycemia. When values are higher than the normal range, it is known as **hyperglycemia** and when the values are lower than the normal limits, it is called **hypoglycemia** (in Greek, hyper=more, hypo=less).

$$\left.\begin{array}{r}\textbf{FBS} > 140\text{ mg/dL}\\ \textbf{PLBS} > 200\text{ g/dL}\end{array}\right\}\text{Indicative of diabetes mellitus}$$

Certain laboratories express glucose concentration in SI units (millimol/Litre). The values in mmol/L can be derived through the following formula.

$$\text{mmol/L} = \frac{\text{Concentration in mg/dL}}{\text{Molecular in weight}} \times 10$$

(Molecular weight of glucose is 180.)

Hyperglycemia is observed in:

1. Most commonly in diabetes mellitus
2. Hyper function of the pituitary and adrenal glands (acromegaly, Cushing syndrome, pheochromocytoma, etc.)
3. Emotional states like fear, anger and anxiety lead to elevation in blood sugar levels because of secretion of epinephrine which has a hyperglycemic action
4. Meningitis, encephalitis, shock
5. Severe hemorrhage
6. Nephrotic syndrome, brain tumour and sepsis

Hypoglycemia is seen in:

1. Insulin overdose in patients of diabetes
2. Pancreatic tumours affecting beta cells
3. Severe liver disease
4. Von Gierke disease
5. Functional or reactive hypoglycemia

Viva Questions

1. What are fasting, postprandial and random blood sugar (FBS, PPBS and RBS) samples? What are their normal glucose levels?
2. What are the blood samples that can be used for glucose estimation? Why is plasma glucose concentration more than that of whole blood?

3. Why should we add an anticoagulant when glucose is estimated in a whole blood sample?
4. What are the methods by which blood glucose can be measured in the laboratory? Which method have you used for glucose estimation? What is the principle of this procedure?
5. How much glucose is present in 1 pint of 5% dextrose saline that is used as IV infusion in patients?
6. What are the causes of hyperglycemia?
7. What are the causes of hypoglycemia?
8. How is the blood glucose level regulated?
9. What are the tests for diagnosing diabetes mellitus?
10. What is the blood glucose level of a newborn infant?

7

BI 11.17

Glucose Tolerance Test

When signs and symptoms in a person are suggestive of **diabetes mellitus** and the blood glucose values are borderline, a glucose tolerance test (GTT) is performed either to confirm or to exclude diabetes. The ability of a person to utilise a given quantity of glucose is known as **glucose tolerance**.

GTT checks the body's ability to process glucose. During the test, glucose levels in blood and urine are monitored for almost three hours after the person has consumed a large dose of sugar solution. In fact, GTT is not based on physiological conditions, because in daily life, such a large quantity of pure glucose is never consumed. Still, GTT is a properly standardised test and hence considered to be the reliable investigation in diagnosing diabetes mellitus in doubtful cases.

Indications for GTT

GTT is required in the following instances:

1. Persons with clinical signs and symptoms suggestive of **diabetes**, but whose blood glucose values are borderline.
2. To determine the prediabetic stage in persons belonging to the risk group (members of a diabetic family)
3. Unexplained glycosuria
4. Diagnosis of GDM (gestational diabetes mellitus)
5. In pregnancy with a history of a large baby (more than 4 kg), abortions
6. To establish IGT (impaired glucose tolerance)
7. Diagnosis of acromegaly

There is no role for GTT in a person who is already diagnosed as diabetic. If a patient shows fasting blood sugar level >140 mg/dL and two-hour postprandial level >200 mg/dL, there is no need for GTT as these values are confirmative of diabetes. When a patient's fasting levels are <100 mg/dL and post-lunch levels <140 mg/dL, the data is sufficient to rule out diagnosis of diabetes and the person is considered as normal. Hence GTT is not regularly required for the diagnosis of diabetes and is not recommended for routine clinical use. It is indicated only when the FBS and PLBS values are inconclusive.

Preparation of the patient for the test

- The patient is asked to eat a high-carbohydrate diet (150–200 g carbohydrate) for 3 days prior to the test. *(The simple way of doing this is by adding 12–15 extra slices of bread to the regular diet each day).* The patient may not be able to tolerate the glucose load during the test. Hence by asking him to take a carbohydrate-rich diet prior to the test, he will be able to tolerate the pure glucose given to him during the test and also have sufficient quantity of glycogen in the liver and other tissues, which is essential for optimum response.

- The patient should be asked to discontinue drugs known to alter blood glucose levels (for example, oral hypoglycemic drugs and insulin injections) for 2 days prior to the test.
- He should avoid strenuous exercise on the previous day as it is likely to increase the rate of oxidation of glucose.
- He should be instructed not to smoke or consume alcohol 12 hours before and during the test.
- He is asked to observe an overnight fast, that is, he should not eat or drink any food or beverages after 8 PM the previous night and he should arrive for the test without taking any breakfast.

Procedure

The patient comes to the laboratory in the morning in the fasting state. He/she must be in a relaxed mood and made to sit comfortably. This is important because if he is anxious or fearful (of the possibility of being a diabetic), it may give false positive results.

Obtain a fasting blood sample and send it for estimation of blood glucose. A corresponding urinary specimen is collected and tested for glycosuria. After collecting these samples, the patient is given a dose of sugar solution (usually 75 g anhydrous glucose dissolved in about 300 mL of water; in case of children 1.75 g/kg body weight). To prevent vomiting, the patient is asked to drink it slowly within five minutes. In the classical procedure, blood and corresponding urine samples are collected at 30-minute intervals for the next 2.5 hours (a total of 6 samples, including the fasting sample).

Glucose is estimated in all six blood samples and the urine specimens are tested qualitatively (Benedict's test) for glycosuria. This is the procedure for **oral glucose tolerance test (OGTT)**. The patient is encouraged to drink water throughout the test procedure to promote adequate urination. He can lie down if he feels faint from the needle punctures.

Intravenous GTT

In cases of malabsorption, nausea and vomiting, GTT can be performed by injecting glucose (0.5 g/kg body weight up to a maximum of 35 g) dissolved in 100 mL of sterile distilled water intravenously within three minutes. Blood samples are collected at 10-minute intervals for the next hour.

Glucose Tolerance Curve

Plot the obtained values on a graph: plasma glucose concentrations are plotted (vertical axis) against the time of collection of samples (horizontal axis) to obtain a glucose tolerance curve (Fig. 7.1).

As the procedure of OGTT is cumbersome and inconvenient to the patient, generally a modified version is usually performed in which the fasting and two-hour post-glucose load samples of blood (only two samples) with corresponding urine specimens are collected and tested. This is sufficient for an optimum result. This is sometimes referred to as **mini-GTT.**

Always watch for any signs of weakness, giddiness, restlessness, nervousness, hunger or sweating (which indicate hypoglycemia) throughout the test.

Interpretation

Normal response

In a normal person, the fasting plasma sugar will usually be 60–90 mg/dL and will never cross 110 mg/dL. After the glucose load, the blood glucose level starts rising and reaches a peak

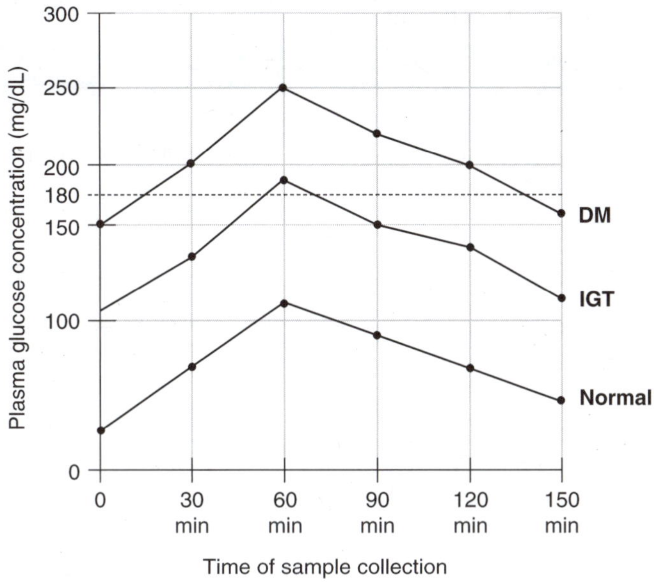

Fig. 7.1 ***Glucose tolerance curves*** *(the dotted line represents renal threshold for glucose)*

Benedict's test for urine glucose						
Normal	–ve	–ve	–ve	–ve	–ve	–ve
IGT	–ve	–ve	+ve	–ve	–ve	–ve
DM	–ve	+ve	+ve	+ve	+ve	–ve

(always less than 140 mg/dL) at the end of one hour and then comes down to normal fasting levels by 2.5–3.0 hours (Fig. 7.1). The initial rise is observed because the quantity of glucose absorbed in the gut exceeds the capacity of the liver and other tissues to use it. However, elevated glucose levels in blood stimulate the secretion of insulin which brings the blood glucose levels back to normal. All the corresponding urine samples give a negative reaction to Benedict's test, indicating that there is no sugar in the urine. Glucose appears in urine (glycosuria) only when its concentration in blood exceeds the renal threshold (usually 180 mg/dL). However, it should be borne in mind that renal threshold is highly variable among individuals (see Tables 7.1 and 7.2).

Table 7.1 *Criteria for the diagnosis of diabetes mellitus (as recommended by WHO)*

	Fasting sample	Two-hour post-load
Normal	<110 mg/dL	<140 mg/dL
Diabetes mellitus	>126 mg/dL	>200 mg/dL
Impaired glucose tolerance (IGT)	110–125 mg/dL	140–199 mg/dL

Table 7.2 *Different types of oral GTT responses*

Different conditions	0 hr	½ hr	1 hr	1½ hr	2 hr
Normal					
Blood glucose (mg/dL)	90	135	145	100	70
Urinary glucose	–	–	–	–	–
Diabetes mellitus					
Blood glucose (mg/dL)	170	210	270	240	220
Urinary glucose	–	+	+++	++	++
Impaired glucose tolerance					
Blood glucose (mg/dL)	115	145	180	170	160
Urinary glucose	–	–	Traces	–	–
Renal glycosuria					
Blood glucose (mg/dL)	85	130	140	105	95
Urinary glucose	–	+	+	+	–

Diabetes mellitus

In diabetes, decreased glucose tolerance is observed. Fasting glucose values are always more than 126 mg/dL and the two-hour post-load values are beyond 200 mg/dL (Table 7.1). The increase in blood glucose level after glucose load is always much greater than in normal persons. The high blood glucose levels are persistent even after 2 hours. In mild diabetes, at least one of the urine specimens (usually 60- or 90-minutes specimen) gives a positive Benedict's test. In severe cases, all urine samples show evidence of the presence of glucose. Ketone bodies are detected in urine samples (Rothera's test is positive) of patients suffering from severe uncontrolled diabetes mellitus.

Diminished glucose tolerance is also seen in hyperpituitarism, hyperthyroidism, Cushing syndrome and severe liver diseases. The GTT curve in these conditions is more or less similar to that of moderate diabetes.

Flat GTT curve

Sometimes, the blood glucose level does not rise as usual following a glucose meal. The curve is flatter than a normal response curve. It may be found in completely normal persons. Such type of increased glucose tolerance is found in malabsorption, hypothyroidism, hypoadrenalism (Addison's disease), hyperinsulinism and hypopituitarism.

Impaired glucose tolerance (IGT)

When the blood glucose levels are above normal but well below diabetic levels, a different type of curve is observed (Fig. 7.1). Such an intermediary glycemic status is known as **impaired glucose tolerance.** In IGT, the FBS level ranges 110–125 mg/dL and the PPBS ranges 140–199 mg/dL (Table 7.1). These persons, though slightly hyperglycemic, are normal. But they need to take certain precautions like restricting carbohydrate content in the diet and exercising regularly. They are at risk of developing frank diabetes in the future if their lifestyle remains unchanged.

In certain cases, only FBS is above normal (110–125 mg/dL) but the two-hour post-glucose value is within the normal range. Such a condition is referred to as **impaired fasting glucose**

or fasting hyperglycemia. As with impaired glucose tolerance, persons with impaired fasting glucose are at increased risk of developing diabetes and cardiovascular disease.

Renal glycosuria

Glycosuria occurs when the glucose concentration in the blood rises beyond the renal threshold level (usually 180 mg/dL). However, in certain individuals, glycosuria is seen even when the glucose concentration in blood is well within the normal limits. This is due to the individual variation of renal threshold. In persons with low renal threshold, glucose appears in urine and such a condition is known as **renal glycosuria.** In such cases, glomerular filtration rate (GFR) is normal but tubular reabsorption is reduced.

Renal threshold is reduced in normal pregnancy and hence renal glycosuria is common during pregnancy. It is a harmless condition and requires no treatment. However, it may be seen in renal tubular transport defects such as '**Fanconi syndrome**'.

Alimentary glycosuria

In some cases, the fasting and two-hour glucose values are normal but a sudden rise in glucose level following the ingestion of glucose is observed. This is due to the very high rate of absorption of glucose from the intestine. One or two urine samples (the one-hour sample) may show glycosuria. This condition is quite normal and may be due to diminished carbohydrate intake for few days. It is also seen in cases of hyperthyroidism and after gastrectomy.

Gestational diabetes mellitus (GDM)

This term represents carbohydrate intolerance identified during pregnancy. It should be noted that diabetic women who become pregnant are not included in this category. **GDM** is characterised by hyperglycemia. In most cases, **GDM** disappears after child birth. However, women with **GDM** are at increased risk of developing diabetes later if proper precautions are not taken.

High glucose levels during pregnancy result in increased secretion of insulin, causing stimulation of fetal growth and increased birth weight of babies. After child birth, the glucose levels have to be measured in the mother to rule out diabetes.

Factors affecting GTT

1. Age: Glucose tolerance is lower in old age
2. Carbohydrate content in the diet
3. Insulin level
4. Physical exercise

The curve is elevated and prolonged in acute infections, liver diseases and hyperthyroidism.

(The reader is advised to refer to Appendix-III "Problem Solving Exercises, Interpretation of Laboratory Data and Case Studies" for interesting exercises.)

Viva Questions

1. What is GTT? When is it performed?
2. What are the factors that affect GTT? What instructions do you need to give to the patient to prepare him/her for the test?

3. What is the response of healthy individuals to GTT?
4. What is impaired glucose tolerance (IGT)?
5. What is renal glycosuria and alimentary glycosuria?
6. What are the diagnostic criteria for diabetes mellitus? How do you interpret the GTT in arriving at a diagnosis?
7. What is gestational diabetes mellitus (GDM)? How are pregnant women screened for it?

8

BI 11.21

Blood Urea and BUN

Urea is the end product of protein metabolism. When the body has taken everything useful out of the protein, all that is left is a toxic ammonia-containing waste called **urea**, which is disposed off eventually.

How is urea formed?

The amino acids obtained either from dietary proteins or from body protein breakdown are metabolised in the body. They are used in the synthesis of new proteins and other nitrogen-containing substances. They are also used to generate energy. The liver converts what is left into either glucose (glucogenic amino acids) or fats (ketogenic amino acids). Before this conversion, the amino acids lose their amino groups which are converted to ammonia. As the ammonia is toxic, it combines with carbon dioxide to form urea in the liver which is released into circulation. Because the ammonia portion of urea is highly toxic, especially to the brain, it must be removed from the body as quickly as possible. The urea is carried to the kidneys from where it is excreted in urine as a waste product.

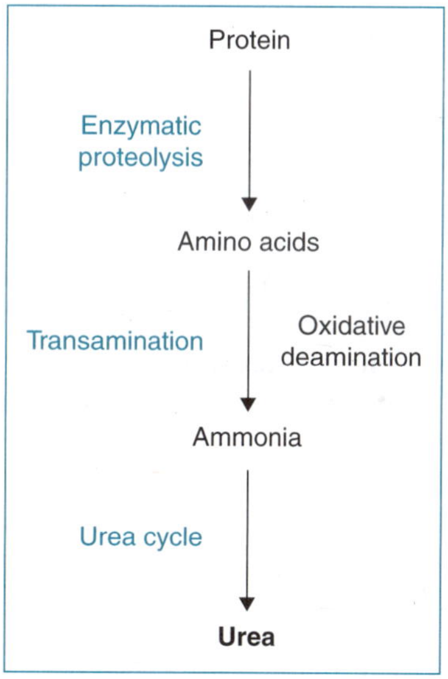

Urea levels in blood

Estimating the urea levels in blood is chiefly concerned with assessing renal function. Urea is the major non-protein nitrogenous (NPN) substance and it represents about 45–50 percent of total NPN of blood and up to 80–90 percent of the total urinary nitrogen excretion.

The normal blood urea level ranges 15–40 mg/dL. It is slightly higher in men than in women. The protein content in the diet will determine the blood urea concentration. However, changes in the protein content in diet will cause only marginal fluctuations in blood urea, and it will be well within the normal limits. The urea levels increase as age advances. The urea levels in young individuals are close to the lower range and the upper limit is normal for elderly people. Hence, a value of 35–40 mg/dL in a person of 25 years should be viewed as suspicious, while the same value in an octogenarian is perfectly acceptable.

Why do urea levels increase with age?

The glomerular filtration rate (GFR) decreases with age. (This is more so after the age of 40). The rate of reabsorption of urea changes inversely with tubular flow. As a result, blood urea levels increase as age advances.

What is BUN?

The molecular weight of urea is 60 and about half of it (28) is contributed by its two nitrogen atoms.

In some laboratories (mostly in the US and Europe), blood urea is represented as **'blood urea nitrogen'** (BUN). Hence, it is necessary for the student to remember the conversion factors for urea-to-urea nitrogen. 60 mg (1 g molecular weight) of urea contains 28 mg (2 g atomic weight) of nitrogen; a factor 0.467 is used to convert urea mass units to those of urea nitrogen, and 2.14 is used to convert urea nitrogen mass units to those of urea.

$$\text{BUN (mg/dL)} = \text{Urea} \times 0.467$$

$$\text{Urea (mg/dL)} = \text{BUN} \times 2.14$$

Methods of Blood Urea Estimation

There is little difference whether whole blood, plasma or serum is used for urea estimation. Blood urea can be estimated by enzymatic (that is, indirect) or by chemical (that is, direct) methods.

Enzymatic methods: These methods are based on the hydrolysis of urea with enzyme urease (source: jack beans, horse gram, and so on), followed by a process that quantitates the ammonium ion.

1. Urease nesselerisation (enzymatic colorimetric method)
2. Berthelot's reaction and enzymatic assay with glutamate dehydrogenase (spectrophotometric approach to ammonium quantitation)

Chemical methods: Most chemical methods for urea are based on the **Fearon reaction** in which diacetyl condenses with urea to form coloured substance, diazine, which absorbs strongly at 540 nm (diacetyl monoxime or DAM method). This method may be performed manually or be automated.

Diacetyl Monoxime (DAM) Method for Blood Urea Estimation

Principle

When a blood sample is treated with diacetyl monoxime (**DAM**) under strong acidic conditions, the urea present in the sample combines with diacetyl in the reagent and forms a compound known as diazine, which is yellow in colour. Diazine reacts with thiosemicarbazide (TSC) and ferric ions and turns pink. The intensity of the pink colour is directly proportional to the concentration of urea in the sample. The intensity of colour is measured using a blue filter (480 nm) in a colorimeter and the concentration of urea is calculated.

Reagents used

1. **10% sodium tungstate**: Dissolve 10 g of sodium tungstate in water and make the volume up to 100 mL with distilled water.
2. **2/3 N sulphuric acid**: Prepared by dilution from a stock solution of 1 N H_2SO_4.
3. **Diacetyl monoxime (DAM) reagent**: Dissolve 2 g of diacetyl monoxime in about 60 mL of water. Add 2 mL glacial acetic acid, dissolve by shaking (warm if necessary) and make the volume up to 100 mL with distilled water.

4. **Sulphuric phosphoric acid reagent (acid mixture)**: Add 150 mL of 85% phosphoric acid to 140 mL water. Mix well. Carefully add 50 mL of concentrated H_2SO_4 slowly while mixing.
5. **Thiosemicarbazide**: 5 g in 1 L of water.
6. **Stock standard solution:** 200 mg urea is dissolved in 100 mL of distilled water (1 mL of stock standard solution contains 2 mg of urea).
7. **Working standard solution:** Dilute 1 mL of stock standard solution with 100 mL of distilled water (1 mL of working standard solution contains 0.02 mg of urea).

Procedure

1. Take three test tubes and mark them as **T (test), S (standard)** and **B (blank)**.
2. Pipette the following into a centrifuge tube: 0.2 mL of blood, 3.4 mL of distilled water, 0.2 mL of 10% sodium tungstate, and 0.2 mL of 2/3 N H_2SO_4.
3. Mix well and centrifuge. This is to remove the proteins.
4. Pipette 2 mL of supernatant into the tube marked 'T'.
5. Take the following solutions in the three test tubes (marked T, S and B) as shown in the table below.

	B	S	T
Supernatant	–	–	2.0 mL
Working urea standard	–	2.0 mL	–
Distilled water	2.0 mL	–	–
Diacetyl monoxime reagent	0.4 mL	0.4 mL	0.4 mL
$H_2SO_4 – H_3PO_4$ mixture	1.6 mL	1.6 mL	1.6 mL
Thiosemicarbazide	1.0 mL	1.0 mL	1.0 mL

6. Mix well and place the test tubes in boiling water for 30 minutes.
7. Cool under running tap water. Read at 480 nm (blue filter).

Note: *If the colour of the test is stronger than the standard, the test is repeated with a diluted sample, after an appropriate dilution.*

	B	S	T
Optical density (OD)			

Calculation

Concentration of urea (mg/dL)

$$= \frac{\text{OD of test} - \text{OD of blank}}{\text{OD of standard} - \text{OD of blank}} \times \frac{\text{Concentration of standard}}{\text{Volume of test}} \times 100$$

(effective volume of blood used)

Note that 2 mL of diluted sample contains 0.1 mL of the original serum. Thus, 2 mL of the dilute standard contains 0.04 mg.

$$= \frac{T}{S} \times \frac{0.04}{0.1} \times 100$$

Thus, concentration of urea (mg/dL) $= \frac{T}{S} \times 40$.

For example, OD of test = 0.60; OD of standard = 1.00.

$$\text{Concentration of urea (mg/dL)} = \frac{0.60}{1.00} \times 40 = 24 \text{ mg/dL.}$$

Clinical interpretation

Blood urea levels are estimated to assess kidney function. Elevated blood urea levels occur in kidney disease, urinary tract obstruction and in conditions that cause protein breakdown, such as burns. Decreased urea levels are observed in malnutrition, severe liver damage and overhydration.

The term '**uremia**' is used to refer to elevated blood urea levels due to renal pathology. Normal blood urea level is 15–40 mg/dL.

The conditions in which blood urea levels are increased can be classified as

1. Pre-renal conditions
2. Renal disease
3. Post-renal causes

Pre-renal

This is associated with increased protein breakdown.

- Vomiting, diarrhea – resulting in dehydration
- Prolonged fever
- Diabetic coma
- Thyrotoxicosis
- Severe burns

Renal

Blood urea is increased in kidney diseases.

- Acute and chronic glomerulonephritis
- Nephrotic syndrome
- Pyelonephritis
- Malignant hypertension
- Renal failure

Post-renal

When there is an obstruction to the flow of urine, the glomerular filtration rate is decreased, with a corresponding increase in blood urea.

- Stones in the urinary tract
- Stricture urethra
- Enlargement of prostate
- Tumour of the bladder

Urea clearance

It is defined as the volume of plasma (or whole blood) that would be completely cleared of urea by excretion of urine in one minute.

Urinary Urea and Urea Clearance

Urinary urea can be estimated similar to blood urea, but the urine must be diluted 1 to 10 with distilled water. The total daily excretion of urine urea in a normal adult is about 25–35 g. The excretion varies with dietary intake of protein.

An elevated urea concentration in urine reflects the concentrating ability of the kidneys. In acute renal failure, urea excretion falls and blood urea levels rise.

When urea concentrations in blood and urine and the volume of urine are known, the urea clearance can be calculated.

$$Cm = \frac{U \times V}{P}$$

where

Cm = maximum urea clearance (mL/minute)
U = urea concentration in urine (mg/mL)
V = volume of urine excreted in mL per minute (estimated by averaging the urine volume collected over 24 hours)
P = urea concentration in plasma (mg/mL).

Normal urea clearance is around 75 mL/minute.

Clinical significance

In the early stages of kidney damage, urea clearance can be used as an indicator of renal damage due to decreased GFR. But in the advanced stages of renal damage, clearance is of little value. Estimating blood urea levels serves as a better indicator of advanced renal disease.

Viva Questions

1. What is urea? How is it formed in the body?
2. What is the normal blood urea level? How much urea is excreted in urine daily?
3. What is BUN?
4. What are the methods by which urea can be estimated in blood?
5. What is the principle of the DAM method?
6. What is the role of 'blank' and 'standard' in quantitative analysis?
7. What is the commonest indication to assess blood urea levels?
8. What are the conditions in which blood urea levels are increased?
9. Name conditions where the urea level in blood is decreased.
10. What is the role of urea clearance in assessing renal damage?

9 Serum and Urinary Creatinine and Creatinine Clearance

BI 11.7, 11.22

Creatinine levels are measured in blood and urine to

- assess kidney function
- screen for kidney damage

Measuring creatinine levels in serum and urine provides a more sensitive measure of kidney damage than other renal parameters. Creatinine is a waste product formed in muscles during creatine metabolism.

Creatine and Creatinine

Creatine constitutes a small fraction of the total non-protein nitrogen of plasma and urine. Creatine is synthesised in the liver, kidneys and pancreas from amino acids (refer to amino acid metabolism for details) and is released into blood and transported to the muscles and the brain, where it is phosphorylated to phosphocreatine, a high-energy compound. The major bulk of creatine is stored as phosphocreatine in skeletal muscle and the interconversion of phosphocreatine and creatine is the driving force of muscle contraction (Lohmann's reaction).

About 2% of creatine is converted spontaneously to creatinine daily and excreted through the kidneys. Thus, the creatinine levels in blood are in proportion to the body's muscle mass. Women and children excrete less creatinine than men because of their smaller muscle mass.

Why is creatinine the best indicator of kidney function?

As creatinine produced from creatine is more or less constant (about 2% of total creatine is converted each day to creatinine), the blood levels of creatinine will not fluctuate much. As the body does not recycle creatinine, it has a constant clearance rate, making it an ideal indicator of kidney function.

Estimation of Creatinine in Serum and Urine

Both chemical and enzymatic methods are used to measure creatinine in body fluids.

Chemical methods: They are based on the reaction with alkaline picrate. Creatinine reacts with picrate ion in an alkaline medium to yield an orange-red complex **(Jaffe's reaction)** which is measured colorimetrically.

Enzymatic methods: Various enzymes of different metabolic pathways such as creatinine deaminase, creatininase and creatinase are involved in the enzymatic measurement of creatinine.

Estimation of Creatinine in Serum by Jaffe's Method

Principle: Creatinine in an alkaline medium reacts with picric acid to form an orange-red tautomer of creatinine picrate, the intensity of which is measured at 520 nm.

Reagents

1. Sodium tungstate, 5% solution
2. H_2SO_4 (2/3 N)
3. NaOH (0.75 N)
4. Picric acid 0.04 M (9.16 g/L)
5. Creatinine stock standard 100 mg/dL (1 mg/mL). Dissolve 1 g of dry creatinine in 1 L of 0.1 N HCl.
6. Working standard (0.04 mg/mL). Dilute 4 mL of stock solution and make it up to 100 mL with water.

Procedure

1. Take three test tubes and labels them as **B (blank), S (standard)** and **T (test)**.
2. **For test**: Dilute 2 mL of serum (or plasma) with 2 mL of distilled water in a centrifuge tube. Add 2 mL of sodium tungstate and 2 mL of 2/3 N H_2SO_4, drop by drop, with constant shaking. Let it stand for 10 minutes. The proteins are precipitated. Filter. Take 3 mL of protein-free filtrate (equivalent to 0.75 mL of the original serum) in the 'T' test tube (Table 9.1).
3. **For standard**: Take 1 mL of the working standard. Add 3 mL of distilled water, 2 mL of sodium tungstate and 2 mL of H_2SO_4. Take 3 mL from this mixture (equivalent of 0.015 mg of creatinine) into the 'S' test tube (Table 9.1).
4. **For blank**: Take 4 mL of distilled water. Add 2 mL of sodium tungstate and 2 mL of H_2SO_4. From this, take 3 mL into the 'B' test tube (Table 9.1).
5. To the above solutions, add 1 mL of picric acid and 1 mL of NaOH.
6. Incubate the tubes for 15 minutes and then read within 30 minutes in a colorimeter at wavelength 520 nm.

Normal values of serum creatinine

Adult males: 0.8–1.5 mg/dL or 65–130 Pmol/L
Adult females: 0.6–1.3 mg/dL or 55–115 Pmol/L
Children: 0.4–1.2 mg/dL or 35–105 Pmol/L

Table 9.1 *Estimation of creatinine in serum by Jaffe's method*

	Blank (B)	Standard (S)	Test (T)
Supernatant	–	–	3 mL
Working creatinine standard	–	3 mL	–
Distilled water	3 mL	–	–
NaOH (0.75 N)	1 mL	1 mL	1 mL
Picric acid	1 mL	1 mL	1 mL

Estimation of creatine: Creatine is converted to creatinine by heating with picric acid in a boiling water bath at 45°C. The creatinine concentration should be estimated twice (before and after heating). The difference is the newly formed creatinine. One gram of creatinine is formed from 1.16 g of creatine. Using this, creatine levels can be calculated.

Normal value of serum creatine: 0.2–0.6 mg/dL

Clinical interpretation

Kidney reserve is large and half the kidney function must be lost before the creatinine level in blood is raised. However, the creatinine levels are taken as more reliable indicators of kidney function than blood urea or uric acid. This is because, unlike other substances, creatinine level is unaffected by diet, exercise or hormonal factors.

A creatinine level >1.5 mg/dL indicates **impaired renal function**. The pre-renal and post-renal conditions which are discussed along with blood urea and BUN are equally applicable for serum creatinine levels.

Calculation

Concentration of creatinine in serum (mg/100 mL)

$$= \frac{\text{OD of test} - \text{OD of blank}}{\text{OD of standard} - \text{OD of blank}} \times \frac{\text{Concentration of standard}}{\text{Volume of test (mL of serum taken)}} \times 100$$

$$= \frac{T}{S} \times \frac{0.015}{0.75} \times 100.$$

$$\text{Concentration of creatinine in serum (mg/dL)} = \frac{T}{S} \times 2.$$

Note that 3 mL of standard solution containing 0.015 mL of creatinine and 3 mL of protein-free filtrate supernatant is equivalent to 0.75 mL of the original serum.

Urine Creatinine

The procedure for serum creatinine can be applied to urinary creatinine as well. The urine should be diluted with water (50 or 100 times) before estimation. The result should be multiplied by the dilution factor.

Procedure

1. Take three test tubes and labels them as **B (blank), S (standard)** and **T (test)**. Add the reagents into the three test tubes as shown in Table 9.2.
2. Creatinine stock standard 100 mg/dL (1 mg/mL).
3. Creatinine working standard 1 mg/dL (0.01 mg/mL). Dilute 1 mL of stock solution and make it up to 100 mL with water.
4. Dilute 1.0 mL urine to 100 mL with water in a standard flask. 3 mL of the diluted urine contains 0.03 mL of urine.

Table 9.2 *Estimation of urine creatinine*

	Blank (B)	Standard (S)	Test (T)
Diluted urine	–	–	3 mL
Working creatinine standard	–	3 mL	–
Distilled water	3 mL	–	–
NaOH (0.75 N)	1 mL	1 mL	1 mL
Picric acid	1 mL	1 mL	1 mL

Calculation

Concentration of creatinine in urine (mg/100 mL)

$$= \frac{\text{OD of test} - \text{OD of blank}}{\text{OD of standard} - \text{OD of blank}} \times \frac{\text{Concentration of standard}}{\text{Volume of test}} \times 100$$

$$= \frac{T}{S} \times \frac{0.03}{0.03} \times 100$$

Note that 3 mL of creatinine working standard contains 0.03 mg of creatinine and 3 mL of diluted urine contains 0.03 mL of urine sample as it is diluted 100 times.

$$= \frac{T}{S} \times 100 \text{ mg/dL.}$$

However, the daily urine output of a healthy individual is 1500 mL.

For 100 mL, concentration of urine creatinine $= \frac{T}{S} \times 100$ mg/dL.

Therefore, for 1500 mL, urine creatinine $= \frac{1500 \times T/S \times 100}{100} = 15 \times \frac{T}{S} \times 100$ mg/day

$$= \frac{15 \times T/S \times 100}{1000} \text{ g/day} = \frac{T}{S} \times 1.5 \text{ g/day} \qquad \text{(converting mg into g).}$$

Normal values for urine creatinine

Men 1.5–2.5 g/day
Women 1.0–2.2 g/day

When the concentration of creatinine is expressed in mg/kg body-weight/24 hours, the values are:

Men: 20–28 Women: 16–21

Urine creatine: The creatinine present before heating is subtracted from the total creatinine present after heating, and the number is converted into creatine by multiplying with a factor of 1.6.

Creatinine coefficient: It is the amount of creatinine and creatine (in mg) put together excreted per kg body-weight per day.

Normal range

Men 24–26 Women 20–22

Clinical interpretation

In normal individuals, creatine excretion is negligible. Increased output of creatine is known as **creatinuria.** Creatinuria is observed in **muscular dystrophies**. Mild creatinuria may be seen in conditions which cause some muscle destruction such as hyperthyroidism, starvation and crush injury.

Creatinine Clearance

It is defined as the volume of plasma (or whole blood) that would be completely cleared of creatinine by excretion of urine in one minute. Creatinine clearance test is a sensitive indicator of renal function. The creatinine level in blood and urine are measured. Volume of urine is determined for a specific time period. The creatinine clearance can be calculated using the following formula.

$$\text{Creatinine clearance} = \frac{U \times V}{S}$$

where U = urine creatinine concentration, S = serum creatinine concentration, V = urine flow in mL/minute.

Normal values

Men 85–125 mL/minute Women 80–115 mL/minute Children 100 mL/minute.

Viva Questions

1. What is creatinine? What are the indications for its measurement?
2. What is the method used to estimate serum and urinary creatinine?
3. What is the normal serum creatinine level? How much creatinine is excreted in urine daily?
4. Why is creatinine estimation considered a better indicator than urea estimation to assess kidney function?
5. What is the difference between creatine and creatinine?
6. How is creatinine synthesised in the body? What is *Lohmann's reaction*?
7. What is the clinical significance of estimating serum creatinine levels?
8. Name the conditions with increased urinary creatinine levels.
9. What is creatinine clearance?
10. What is creatinine coefficient?

10

BI 6.2; 6.3; 6.4; 11.17

Uric Acid

Uric acid is the end product of the catabolism of purine nucleotides. Its levels are measured to detect gout and kidney dysfunction. Purine nucleotides are the main components of nucleic acids. The total body uric acid pool which is about 1200 mg in normal men and 600 mg in women shoots up to as high as 20,000 to 30,000 mg in patients of *gouty arthritis*.

Formation and Excretion of Uric Acid

Purines are obtained from three sources in the body (Fig. 10.1): diet, degradation of endogenous nucleotides and de novo synthesis. The main dietary sources include organ meats such as the liver, heart, kidney, poultry and fish. Since uric acid is the end metabolite of purines, the concentration of uric acid in plasma depends on the catabolic rate of purines and its excretion from the body.

The major chunk of uric acid (more than 75%) is excreted into urine. The remainder is secreted into the gut and converted to allantoin, carbon dioxide and ammonia by bacterial action. This process is known as 'uricolysis'.

Uric acid and the kidneys

In the kidneys, uric acid is filtered completely at the glomerulus and is fully reabsorbed in the proximal convoluted tubule. Distally, both secretion and reabsorption occur. As a net result, 6–12 percent of the filtered uric acid is finally excreted into urine.

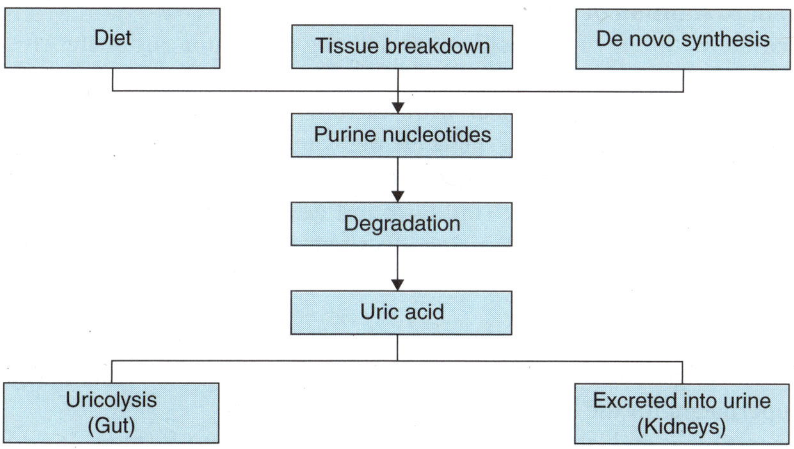

Fig. 10.1 *Formation and excretion of uric acid*

Methods of Estimation

1. **Caraway's method**: This method is based on the formation of 'tungsten blue' when phosphotungstic acid is reduced by urate in an alkaline medium.
2. **Uricase methods**: These methods are more specific than the phosphotungstic acid method. Enzymatic conversion of urate to allantoin is the underlying mechanism in these methods. Today, various devices that utilise the enzyme uricase in a dry reagent format are used to measure uric acid concentration in body fluids.
3. **HPLC** (high pressure liquid chromatography) methods.

Estimation of Serum Uric Acid by Caraway's Method

Principle: Phosphotungstic acid in alkaline medium oxidises uric acid to allantoin and carbon dioxide and itself gets reduced to tungsten blue which is measured colorimetrically at 700 nm.

Reagents

1. Sodium tungstate (10%)
2. H_2SO_4 (2/3 N)
3. **Tungstic acid**
   (a) Add 50 mL of 10% sodium tungstate, 50 mL of 2/3 N H_2SO_4 and a drop of phosphoric acid to 800 mL of distilled water and mix.
   (b) Discard when cloudy. Store in a brown bottle.
4. **Phosphotungstate reagent**
   (a) Dissolve 50 g sodium tungstate in about 400 mL of distilled water.
   (b) Add 40 mL of 85% phosphoric acid. Boil gently under reflux condenser for two hours.
   (c) Cool and make the volume up to 500 mL.
   (d) Store in a brown bottle to protect it from light. The reagent is stable for about two years. Dilute 1 to 10 for use.
5. Sodium carbonate 10%
6. **Stock standard solution of uric acid (100 mg/dL)**
   Weigh accurately 100 mg of uric acid AR and 60 mg of lithium carbonate AR (Li_2CO_3) and add them to a beaker. Add a few drops of distilled water and warm the contents gently to dissolve the solids. Cool and transfer with washing to a volumetric flask and make up to 100 mL with distilled water. This is stable for many months if refrigerated.
7. **Working standard of uric acid solution (0.005 mg/mL)**
   Pipette 0.5 ml of uric acid stock standard and add it to a 100 mL standard flask and make the volume up to 100 mL using distilled water. It is stable for 2 to 3 days if refrigerated.

Procedure

1. Take 5.4 mL of tungstic acid in a centrifuge tube and add 0.6 mL of serum to it and centrifuge. Use the supernatant for the test.
2. Take three test tubes and label them **T (test), S (standard)** and **B (blank).** Add the reagents into the three test tubes as shown below.

	(B)	Standard (S)	Test (T)
Distilled water	3.0 mL	–	–
Uric acid working	–	3.0 mL	–
Standard (0.84 mg/100 mL)			
Supernatant	–	–	3.0 mL
Sodium carbonate	0.6 mL	0.6 mL	0.6 mL
Phosphotungstate	0.6 mL	0.6 mL	0.6 mL

3. Mix well and keep them at room temperature for 30 minutes.
4. Take the reading at 700 nm or by using a red filter within 15 minutes of adjusting the blank.

Calculations

The supernatant fluid used in the test is equal to 0.3 mL of the original serum (volume of the test) and 3 mL of the uric acid; the working standard contains 25.2 mg of urate (concentration of standard).

$$\left.\begin{array}{c}\text{Concentration of uric acid}\\ \text{in serum (mg/100 mL)}\end{array}\right\} = \frac{\text{OD of test}}{\text{OD of standard}} \times \frac{\text{Concentration of standard}}{\text{Volume of test}} \times 100$$

$$= \frac{T}{S} \times \frac{0.015}{0.3} = \frac{T}{S} \times 5 \text{ mg.}$$

Estimation of Uric Acid by Uricase Method

Principle: Enzymatic conversion of uric acid to allantoin produces H_2O_2, measured through an indicator reaction involving peroxidase.

$$\text{Uric acid} + H_2O \xrightarrow{\text{uricase}} \text{Allantoin} + H_2O_2$$

$$H_2O_2 + \text{Phenolic chromogen} \xrightarrow{\text{peroxidase}} \text{Phenolin (a red-coloured compound)}$$

Sample used: Serum/Heparinised plasma/EDTA plasma

Reagents used

1. Buffered enzymes
2. Standard uric acid solution (6 mg/100 mL)

Sample preparation

Sample can be stored for up to three days at 4°C.

Procedure

1. Take three test tubes and label them **test, standard** and **blank**.
2. Add the reagents as shown below.

	Blank	Standard	Test
Distilled water	0.025 mL	–	–
Standard solution of uric acid	–	0.025 mL	–
Sample	–	–	0.025 mL
Buffered enzymes	1.00 mL	1.00 mL	1.00 mL

3. Incubate the tubes at room temperature for 10 minutes.
4. Take OD readings of the sample and standard against the blank.

If 100 mL contains 6 mg, then 0.025 mL will contain 0.0015 mg.

Calculation

$$\left.\begin{array}{l}\text{Concentration of}\\ \text{uric acid (mg/dL)}\end{array}\right\} = \frac{\text{OD of test} - \text{OD of blank}}{\text{OD of standard} - \text{OD of blank}} \times \frac{\text{Concentration of standard}}{\text{Volume of test}} \times 100$$

$$= \frac{\text{OD of test}}{\text{OD of standard}} \times \frac{0.0015}{0.025} \times 100.$$

$$\text{Concentration of uric acid (mg/dL)} = \frac{\text{T}}{\text{S}} \times 6.$$

Hyperuricemia and Gout

Uric acid is the end metabolite of purine metabolism. (*For more details about the metabolic pathways leading to uric acid synthesis, please refer to Textbook of Biochemistry, 4th Edition*).

Hyperuricemia occurs due to increased formation of uric acid, decreased excretion or a combination of both. Some alcoholic beverages contain high amounts of purines and so excessive consumption of alcohol increases nucleotide degradation, resulting in hyperuricemia.

At normal pH, uric acid occurs in ionised form (urate ion). In extracellular fluid (ECF), it occurs in combination with sodium as **monosodium urate**. It is minimally soluble and the ECF is easily saturated with monosodium urate just above the normal level, forming crystals. Monosodium urate crystals are usually deposited in synovium, the synovial fluids of joints, causing severe joint pain and swelling with redness. This inflammatory condition is popularly known as gout **(gouty arthiritis)**.

What happens in the joints?

Monosodium urate crystals in the joints are taken up by polymorphic leucocytes, but these crystals damage the lysosomes of these cells, causing cellular disruption. The lysosomal enzymes in association with interleukins and other inflammatory mediators from monocytes and tissue macrophages coupled with the generation of free radicals (superoxide) cause an **acute inflammatory reaction**. In gouty arthritis, mostly the peripheral joints (for example, the first metatarso phalangeal joint) are involved because they have low intra-articular temperatures (monosodium urate salt is minimally soluble and maximally precipitable at lower temperatures).

Gout is more common is men and post-menopausal women. It rarely occurs in women of the reproductive age group.

Management

Non-steroidal anti-inflammatory drugs (indomethacin) and colchicine are used to relieve symptoms of gout. But these drugs do not affect hyperuricemia. Hyperuricemia is treated with allopurinol. Allopurinol is an inhibitor of xanthine oxidase and thus inhibits the synthesis of uric acid from xanthine. Dietary meat consumption and avoiding alcohol helps.

Pseudogout: The deposition of calcium pyrophosphate in the joints mimics gout in signs and symptoms.

Other conditions associated with hyperuricemia

Apart from gout, the following disorders are associated with hyperuricemia.

1. **Renal calculi**: Uric acid accounts for 10% of renal stones. (The majority of renal stones are composed of calcium oxalate and calcium phosphate).
2. **Tophi**: Deposits of sodium urate in soft tissues.
3. **Lesch–Nyhan syndrome**: Deficiency of the enzyme hypoxanthine guanine phosphoribosyl transferase (HGPRT) results in diminished salvage pathway, decreasing purine reutilisation and causing enhanced uric acid formation.
4. **Von Gierke disease (glycogen storage disease type I)**: Glucose-6-phosphate that accumulates due to deficiency of the enzyme glucose-6-phosphatase is diverted to HMP shunt pathway resulting in increased synthesis of ribose-5-phosphate, which in turn increases the levels of phosphoribosyl pyrophosphate and enhances the metabolism of purine nucleotides to uric acid.
5. Lactic acidosis decreases uric acid secretion in the renal tubules, contributing to elevated urate levels in plasma.

Hypouricemia

Reduced urate levels are uncommon and are of little significance. Hypouricemia may be seen in severe liver disease and renal tubular defects (for example, Fanconi syndrome).

Viva Questions

1. What is uric acid? How is it formed in the body?
2. What is the importance of estimating serum uric acid?
3. What are the methods of serum uric acid estimation?
4. What are the diseases associated with hyperuricemia?
5. How do elevated uric acid levels cause gout?
6. Consumption of alcohol may precipitate acute attack of gout. What is the underlying biochemical mechanism?
7. What are *tophi*?
8. Why is *allpurinol* the drug of choice in the treatment of gout?
9. Does hyperuricemia cause kidney stones? How?
10. Why is hyperuricemia often associated with cancers?

11 Serum Total Proteins and Albumin–Globulin Ratio

BI 11.8; 11.22

Blood is composed of cells such as RBCs, WBCs and platelets suspended in a liquid called plasma. About 55–60 percent of blood is made up of plasma. Plasma consists of water, electrolytes, nutrients, metabolites and various proteins. Plasma proteins include albumin, globulin and fibrinogen. All the plasma proteins except the γ-globulins are synthesised in the liver. Albumin is the main contributor to total plasma protein.

If blood is kept without adding any anticoagulant, after some time it will clot; the remaining fluid portion is called serum. Hence **serum** is plasma minus fibrinogen.

The total concentration of plasma proteins is about 6–8 g/dL. Albumin concentration is 3.5–5.5 g/dL. Total globulins concentration is 2.0–3.0 g/dL.

Albumin–Globulin (A/G) Ratio

The normal A/G ratio is roughly about 1.5:1 to 2:1. The A/G ratio is either lowered or reversed in conditions in which either albumin is decreased or globulins are increased.

Decreased levels of albumin are found in

1. Severe protein malnutrition
2. Cirrhosis of the liver and other liver diseases (synthesis of albumin by the liver is affected)
3. Nephrotic syndrome (due to increased excretion of albumin into urine)

Increased levels of globulins are seen in

1. Multiple myeloma (a plasma cell cancer)
2. Chronic infections

In the above conditions, the A/G ratio is either lowered or completely reversed. Estimation of total serum proteins can be done by a number of methods.

1. Precipitation methods
2. Micro kjeldahl method
3. Colorimetric methods (biuret method)
4. Electrophoretic separation and quantitation by densitometry

Estimation of Total Serum Proteins by Biuret Method

Biuret is the method most commonly used to measure total protein concentrations in serum.

Reagents used

1. **Stock biuret reagent**
   (a) Dissolve 45 g of sodium potassium tartrate in 400 mL of 200 mmol/L NaOH.

(b) Add 15 g of $CuSO_4$, stirring continuously until the solution is complete.
(c) Add 5 mL of potassium iodide and make up the volume to 1 L with 200 mmol/L of NaOH.

2. **Working biuret reagent**
   (a) Dilute 200 mL of stock solution to 1 L with 200 mmol/L of NaOH.
3. **Normal saline**: 0.9% NaCl
4. **Protein standard solution** prepared by dissolving 200 mg of bovine albumin in 100 mL of distilled water.

Principle: Copper present in biuret reacts with peptide bond nitrogen and forms a violet-coloured complex. The intensity of the violet colour is directly proportional to the amount of total protein present in the given sample.

Procedure

1. Label three test tubes as **blank, standard** and **test.**
2. Pipette out 3 mL of distilled water into blank, 3 mL of standard solution into standard and 0.1 mL of the sample plus 2.9 mL of normal saline into test (see table below).
3. Add 3 mL of biuret reagent to all three test tubes and incubate them at room temperature for 10 minutes.
4. After 10 minutes, measure the intensity of the colour in a colorimeter at 535 nm.

	Blank (B)	Standard (S)	Test (T)
Distilled water	3.0 mL	–	–
Standard solution	–	3.0 mL	–
Sample	–	–	0.1 mL
Normal saline	–	–	2.9 mL
Biuret working reagent	3.0 mL	3.0 mL	3.0 mL
Optical density (OD) (at 535 nm)			

Calculation

The amount of total protein present in the given sample of serum is calculated as follows.

$$= \frac{\text{OD of test} - \text{OD of blank}}{\text{OD of standard} - \text{OD of blank}} \times \frac{\text{Concentration of standard}}{\text{Volume of test}} \times 100$$

(effective volume of serum used)

When 100 mL contains 200 mg protein, 3 mL contains 6 mg.

$$= \frac{\text{OD of test} - \text{OD of blank}}{\text{OD of standard} - \text{OD of blank}} \times \frac{6}{0.1} \times 100 \times \frac{1}{1000}\ \text{g/dL}$$

$$= \frac{T}{S} \times 6\ \text{g/dL}$$

Measuring the A/G Ratio

1. Take 5.8 mL of 28% sodium sulphite in a centrifuge tube and add 0.2 mL of undiluted serum. Gently rotate the tube, enabling even distribution of the serum in the solution.
2. Add 2 mL of ether. Place your thumb on the mouth of the tube and gently move it upside down slowly, about 20 times, and keep the contents for half an hour.
3. Centrifuge the contents of the tube for 10 minutes.
4. Using a capillary dropper, carefully detach the globulin portion formed at the junctional point of the water–ether layer. Remove the supernatant and transfer it into a tube labelled 'albumin'.
5. Add the contents, as shown in table below, and keep at room temperature for 10 minutes.
6. Read the OD in a colorimeter at 520 nm.

	Blank	Standard	Test
Supernatant	–	As in total protein	3 mL
Normal saline	3 mL	–	–
Biuret reagent	3 mL	3 mL	3 mL
Optical density (OD) at 520 nm			

Calculation

The concentration of albumin is calculated by the following formula.

$$= \frac{\text{OD of test} - \text{OD of blank}}{\text{OD of standard} - \text{OD of blank}} \times \frac{\text{Concentration of standard}}{\text{Volume of test}} \times 100 \text{ mg/dL}$$

The volume of the test is 0.1 mL (5.8 + 0.2 mL solution contains 0.2 mL of serum; thus, 3 mL will contain 0.1 mL).

Interpretation

Increase in total protein is seen in

1. Multiple myeloma
2. Dehydration
3. Chronic infections

Note that whenever the protein level is increased, it is usually due to increase in globulin concentration but not albumin.

Total proteins are decreased in

1. Malnutrition (kwashiorkor), malabsorption syndromes
2. Nephrotic syndrome
3. Cirrhosis of the liver
4. Severe burns
5. Hemorrhage
6. Tuberculosis in severe form

Estimation of Albumin

Albumin is a simple globular protein synthesized in liver. It is a major plasma protein that helps in regulation and distribution of cellular fluids. It helps in the transport of hormones, metals , calcium, bilirubin and fatty acids.

Albumin is estimated as follow:

1. Take 5.8 mL of 28% sodium sulphite in a centrifuge tube and add 0.2 mL of undiluted serum. Gently rotate the tube, enabling even distribution of the serum in the solution.
2. Add 2 mL of ether. Place your thumb on the mouth of the tube and gently move it upside down slowly, about 20 times, and keep the contents for half an hour.
3. Centrifuge the contents of the tube for 10 minutes.
4. Using a capillary dropper, carefully detach the globulin portion formed at the junctional point of the water–ether layer. Remove the supernatant and transfer it into a tube labelled 'albumin'.
5. Add the contents and keep at room temperature for 10 minutes.
6. Read the OD in a colorimeter at 520 nm table below.

	Blank	Standard	Test
Supernatant	–	As in total protein	3 mL
Normal saline	3 mL	–	–
Biuret reagent	3 mL	3 mL	3 mL
Optical density (OD) at 520 nm			

Calculation

The concentration of albumin is calculated by the following formula.

$$= \frac{\text{OD of test} - \text{OD of blank}}{\text{OD of standard} - \text{OD of blank}} \times \frac{\text{Concentration of standard}}{\text{Volume of test}} \times 100 \text{ mg/dL}$$

The volume of the test is 0.1 mL (5.8 + 0.2 mL solution contains 0.2 mL of serum; thus, 3 mL will contain 0.1 mL).

Serum albumin concentration is 3.5–5.5 g/dL.

Increased levels of albumin are seen in dehydration.

Decreased levels of albumin are found in

1. Severe protein malnutrition, skin burns, tissue damage
2. Cirrhosis of the liver and other liver diseases (synthesis of albumin by the liver is affected)
3. Nephrotic syndrome (due to increased excretion of albumin into urine)

Estimation of Globulins

It is a calculated parameter, calculated as follows:

Globulins = Total Proteins – Albumin

Total globulins concentration is 2.0–3.0 g/dL.

Increased levels of globulins are seen in

1. Multiple myeloma (a plasma cell cancer)
2. Autoimmune disorders
3. Chronic infections

Albumin–Globulin (A/G) Ratio

It is a calculated parameter given by the ratio of albumin to globulin i.e., Albumin/Globulin ratio. The normal A/G ratio is roughly 1.5:1 to 2:1. The A/G ratio is either lowered or reversed in conditions in which either albumin is decreased or globulins are increased.

Clinical significance of A/G ratio

Reversal of A/G ratio is seen in acute/chronic liver failure, malnutrition, acute/chronic kidney failure and nephrotic syndrome.

Viva Questions

1. Name the plasma proteins. What are the normal levels of total serum proteins?
2. What are the functions of albumin and various globulins?
3. How are plasma proteins separated in the laboratory?
4. What is the difference between plasma and serum?
5. What is the method adopted to estimate serum proteins?
6. What is the importance of the 'blank' and 'standard' solutions in quantitative analysis?
7. What is the A/G ratio? How much is normal? What is its significance?
8. Name any two conditions in which the A/G ratio is reversed.
9. Explain the terms hypoproteinemia and hyperproteinemia. Which is more common?
10. Mention the common causes of hypoproteinemia or hypoalbuminemia. What is the common clinical sign of hypoalbuminemia and the biochemical basis for it?
11. What is multiple myeloma? What is the protein that is excreted in the urine of these patients?
12. What are albuminemia and microalbuminuria?

12 Cholesterol and Lipid Profile

BI 11.9

Elevated levels of lipids are always viewed as a warning signal for cardiovascular disease. Elevated levels of serum cholesterol and other lipid fractions are associated with the incidence of atherosclerosis and coronary heart disease. Cholesterol is present exclusively in animal-derived foods. It is an integral component of biological membranes, is essential for nerve impulse conduction and acts as the parent molecule for the biosynthesis of steroids in the body. Cholesterol is present in blood in free form (30%) and esterified form (70%).

Cholesterol levels are estimated in serum to:

- Measure the risk of cardiovascular disease
- Check the body's fat metabolism
- Help in the diagnosis of liver disease, cholestasis (gall stones), pancreatitis, nephrotic syndrome and thyroid diseases

Total cholesterol levels vary with age and sex. The normal range is 150–200 mg/dL (preferably <180 mg/dL). Levels >280 mg/dL are considered abnormal.

Estimation of Serum Cholesterol by Zak's Ferric Chloride Method

Reagents used

1. Ferric chloride, 0.05%
2. Sulphuric acid
3. Glacial acetic acid (aldehyde free)
4. Cholesterol stock standard (100 mg per 100 mL in acetic acid)
5. Cholesterol working standard (0.04 mg per mL). Dilute 4 mL of stock to 100 mL with ferric chloride reagent ($FeCl_3$ in glacial acetic acid).

Principle

Cholesterol in acetic acid reacts with $FeCl_3$ and H_2SO_4 to produce a red colour. The absorbance of the red-coloured solution is read at 560 nm in a colorimeter.

Procedure

Constitute blank, standard and test solutions as given below.

Preparation of supernatant

Reagents	Blank	Standard	Test
Cholesterol working	–	9.9 mL	–
Standard			

Reagents	Blank	Standard	Test
NaCl (0.9%)	–	0.1 mL	–
Serum	–	–	0.1 mL
$FeCl_3$	–	–	9.9 mL

Mix well and keep for 10 minutes and centrifuge.

Final steps of cholesterol estimation

Reagents	Blank	Standard	Test
Supernatant	–	5 mL	5 mL
$FeCl_3$	5 mL	–	–
H_2SO_4	3 mL	3 mL	3 mL

Mix well and incubate for 30 minutes at room temperature. Read absorbance at 560 nm in a colorimeter.

Calculation

Concentration of cholesterol (mg/dL)

$$= \frac{T}{S} \times \text{concentration (mg)} \times \frac{\text{Total solution}}{\text{Solution taken}} \times \frac{\text{100 mL}}{\text{Serum taken}}$$

$$= \frac{T}{S} \times 0.2 \times \frac{10}{5} \times \frac{\text{100 mL}}{0.1}$$

$$= \frac{T}{S} \times \text{400 mg/dL}$$

Interpretation

Increased cholesterol levels (hypercholesterolemia) are observed in

1. Diabetes mellitus
2. Hyperlipoproteinemias types II–IV
3. Nephrotic syndrome
4. Cirrhosis of liver
5. Obstructive jaundice
6. Hypothyroidism (myxedema)
7. Coronary artery disease
8. Xanthomatosis

Decreased cholesterol levels (hypocholesterolemia) are observed in

1. Hyperthyroidism
2. Hemolytic jaundice

3. Malabsorption syndrome
4. Pernicious anemia

Plasma Lipid Profile

HDL-Cholesterol (HDL-C)

HDL (high density lipoprotein) removes cholesterol from various tissues and esterifies it with the help of LCAT enzyme and Apo-A1. It then transfers it to the liver. In the liver, cholesterol is further metabolised and eventually eliminated from the body. In this way, HDL plays an important role in removing the excess cholesterol, thereby performing a role beneficial to the body. This is the reason HDL-cholesterol is popularly known as '*good cholesterol*'.

Principle of estimation

The serum is treated with phosphotungstic acid and magnesium chloride to precipitate LDL, VLDL and chylomicrons. Centrifugation leaves HDL in the supernatant. Cholesterol in the HDL fraction is then measured by the same method used for serum cholesterol estimation.

LDL-Cholesterol (LDL-C)

LDL (low density lipoprotein) delivers cholesterol to various tissues. Hence, when the concentration of LDL-C increases, there is a chance of excess cholesterol deposition in the peripheral tissues. This increases the risk of atherosclerosis and coronary artery disease (CAD). This is the reason LDL-cholesterol is called '*bad cholesterol*'.

Principle of estimation

Unlike HDL-C, which is directly estimated, LDL-C is indirectly calculated from the serum triacylglycerol (TAG) level because the separation of LDL from serum is a cumbersome process.

A formula (*Friedewald equation*) is usually used to estimate LDL-C levels. According to the Friedewald equation, the total cholesterol (TC) is equal to the sum of cholesterol present in HDL, LDL and VLDL.

$$\text{TC} = \text{HDL-C} + \text{LDL-C} + \text{VLDL-C} \text{ (all in mg/dL)}$$

$$\begin{aligned}\text{LDL-C} &= \text{TC} - [\text{HDL-C} + \text{VLDL-C}] \\ &= \text{TC} - [\text{HDL-C} + \text{TAG}/5]\end{aligned}$$

Usually, VLDL-C is taken as one-fifth of the serum triacylglycerol (TAG) level based on the assumption that the ratio of TAG and cholesterol content of VLDL is 5:1. However, this calculation is not applicable when the TAG level exceeds 400 mg/dL or if the patient has type III hyperlipoproteinemia, because these conditions alter the usual composition of VLDL.

HDL-C and LDL-C ratio

The ratio of HDL-C and HDL-C is considered a more valuable marker to assess the risk of atherosclerosis than the individual levels of LDL-C and HDL-C. The normal LDL-C: HDL-C ratio is 2.0–2.5. A high ratio is considered a predictor of risk of CAD.

Interpretation

Normal HDL levels – Men: 30–60 mg/dL; Women: 40–70 mg/dL
Normal LDL levels – 80–130 mg/dL

HDL-C levels are higher with
- Regular physical activity
- Regular exercise

HDL-C levels are lower in
- Malnutrition
- Frequent fasting/starvation
- Lack of physical activity/exercise
- Diabetes mellitus
- Hypothyroidism
- Liver disease
- Chronic smoking
- Usage of anabolic steroids
- Certain rare dislipidemias (for example, Tangier disease)

LDL-C levels are higher in
- Coronary artery disease
- Diabetes mellitus
- Nephritic syndrome
- Cirrhosis of liver
- Hypothyroidism
- Obstructive jaundice

LDL-C levels are lower in
- Hyperthyroidism
- Malabsorption syndrome
- Hemolytic jaundice
- Pernicious disease

Estimation of Triacylglycerols (TAG)

Triacylglycerols (TAG) or triglycerides (TG) are the main component of chylomicrons and VLDL. Their levels in serum reflect the status of lipid metabolism. Elevated TAG levels are not directly associated with coronary heart disease (CHD).

Principle of estimation

TAG is extracted from serum using isopropanol and then saponified by propanolic potassium hydroxide to release glycerol, which is oxidised by periodate to form formaldehyde. Formaldehyde reacts with acetyl acetone to produce a yellow-coloured compound which is measured on a colorimeter at 450 nm. Phospholipids, which are also hydrolysed to glycerol and may interfere with the estimation of TAG, are removed by adsorption on alumina.

Reagents used

Isopropanol, alumina, propanolic potassium hydroxide, sodium metaperodate, acetyl acetone and TAG standard (200 mg/dL)

Procedure

1. Take three test tubes and label them blank (B), standard (S) and test (T).
2. Add reactants as given in below.

Reagents	Blank (B)	Standard (S)	Test (T)
Distilled water	0.2 mL	–	–
Standard	–	0.2 mL	–
Serum	–	–	0.2 mL
Isopropanol	7.8 mL	7.8 mL	7.8 mL
Alumina	1.0 g	1.0 g	1.0 g

3. Mix the contents in all the tubes and then centrifuge.
4. Take another set of three test tubes and label them blank (B), standard (S) and rest (T).
5. Add reactants, as given in below.

Reagents	Blank (B)	Standard (S)	Test (T)
Supernatant (from the 'B' tube)	4.0 mL	–	–
Supernatant (from the 'S' tube)	–	4.0 mL	Supernatant (from the 'B' tube)
Supernatant (from the 'T' tube)	–	–	4.0 mL
Propanolic potassium hydroxide	1.2 mL	1.2 mL	1.2 mL
Incubate at 65°C for 15 minutes			
Metaperiodate	2.0 mL	2.0 mL	2.0 mL
Acetyl acetone	1.0 mL	1.0 mL	1.0 mL

6. Incubate the tubes at 65°C for 30 minutes.
7. Cool the tubes and read the standard and test against blank at 450 nm (violet filter).
8. Take the OD readings in the colorimeter and use them in the following formula.

Calculation

$$= \frac{\text{OD of test} - \text{OD of blank}}{\text{OD of standard} - \text{OD of blank}} \times \frac{\text{Concentration of standard}}{\text{Volume of test}} \times 100$$

$$= \frac{\text{T} - \text{B}}{\text{S} - \text{B}} \times \frac{0.2}{0.1} \times 100$$

$$= \frac{\text{T}}{\text{S}} \times 200 \text{ mg/dL}$$

$$\text{Concentration of TAG} = \frac{\text{T}}{\text{S}} \times 200 \text{ mg in 100 mL of serum.}$$

Interpretation

Normal range of TAG is 50–200 mg/dL.

Serum TAG is higher in

- Diabetes mellitus
- Hypothyroidism
- Chronic pancreatitis
- Alcoholism
- Chronic smoking
- Lack of physical activity
- Steroid medication

Serum TAG is lower in

- Malnutrition
- Malabsorption syndrome
- Hyperthyroidism
- Liver disease
- Abetalipoproteinemia

Viva Questions

1. What is the biological role of cholesterol in the body?
2. Name the important substances synthesised from cholesterol.
3. Name two important complications of hypercholesterolemia.
4. What is the importance of estimating serum cholesterol?
5. What is the normal cholesterol level in blood?
6. What are the common dietary sources of cholesterol?
7. How is cholesterol removed from the body?
8. What is the Libermann–Burchard reaction? What is its significance?
9. What are the various lipoprotein fractions?
10. What is HDL-C? How much is normal?
11. What is LDL-C? How much is normal?
12. What is the clinical significance of HDL-C and LDL-C? Why is HDL-C considered as 'good cholesterol' and LDL-C as 'bad cholesterol'?
13. What is reverse cholesterol transport?
14. What is the significance of the LDL-C:HDL-C ratio?
15. What are the common causes of low HDL-C and high LDL-C?
16. What are the causes of elevated TAG levels?
17. What lifestyle modifications do you advise to maintain a normal lipid profile?

13 Bilirubin

BI 11.12

Bilirubin is a bile pigment and is the end product of heme catabolism. Any increase of bilirubin in blood is indicative of jaundice. It is produced by the break-down of RBCs. It is excreted in urine as urobilinogen and in feces as stercobilinogen.

When is serum bilirubin estimated?

- To evaluate liver function and to diagnose obstruction in the biliary tract.
- To diagnose various types of jaundice and in their follow-up.
- To diagnose anemia due to hemolysis.
- To determine the risk of excessive bilirubin in a newborn. (In newborn babies, excess bilirubin can cross the blood–brain barrier and cause irreversible damage to brain tissue).

Bilirubin exists in two forms: conjugated bilirubin and unconjugated bilirubin. Bilirubin is conjugated in the liver by glucuronate and excreted in bile.

Serum Bilirubin

The normal concentration of total serum bilirubin is 0.2 to 1.0 mg/dL. Of this, the conjugated bilirubin is 0.2 to 0.4 mg/dL and the unconjugated bilirubin is 0.2–0.8 mg/dL.

Estimation of Serum Bilirubin by Van den Bergh Reaction

This is the specific reaction to identify the elevated serum bilirubin above the normal level. Normal serum gives a negative Van den Bergh reaction.

Principle

Bilirubin is estimated by using the diazo reagent. Diazotised sulphanilic acid (sulphanilic acid treated with a mixture of sodium nitrite and HCl) in the presence of bilirubin forms a purple-coloured azobilirubin, which is measured colorimetrically.

Direct and indirect reactions: Two types of reactions can be observed: direct and indirect. Bilirubin is water insoluble but when conjugated with glucuronate it becomes soluble. When the reaction is carried out in an aqueous medium, the **conjugated bilirubin** gives a purple colour very quickly (within 30 seconds). This is called **direct positive** Van den Bergh reaction. Addition of methanol dissolves the **unconjugated bilirubin**, which then answers the Van den Bergh reaction (within 30 minutes) and is called **indirect positive.**

If the serum contains both unconjugated and conjugated bilirubin in high concentrations, the purple colour is produced immediately (direct positive) which is further intensified by the addition of methanol (indirect positive). Methanol disrupts the intramolecular hydrogen bonds, and thus both conjugated and unconjugated bilirubin react in giving the measure of total bilirubin. This type of reaction is known as **biphasic**.

Van den Bergh reaction is useful in differentiating the types of jaundice.

Indirect positive – Hemolytic jaundice
(↑unconjugated bilirubin)

Direct positive – Obstructive jaundice
(↑conjugated bilirubin)

Biphasic – Hepatic jaundice
(↑both types of bilirubin)

Estimation of Serum Bilirubin by Malloy and Evelyn Method

Van den Bergh reaction was adapted by Malloy and Evelyn to develop a method for the estimation of serum bilirubin.

Reagents used

1. *Diazo reagent*

   Diazo-A: Weigh 0.1 g of sulphanilic acid and dissolve it in 1.5 mL of concentrated HCl. Make it up to 100 mL with distilled water.

   Diazo-B: Weigh 0.5 g of sodium nitrite and dissolve in distilled water to make the volume up to 100 mL.

   The diazo reagent is prepared fresh by adding 0.3 mL of Diazo-B to 10 mL of Diazo-A.
2. *Absolute methanol*
3. *HCl (1.5%):* 1.5 mL of concentrated HCl is diluted to 35 mL with distilled water.
4. *Bilirubin standard solution:* Prepare 10 mg/dL bilirubin in chloroform.

Procedure

1. Dilute 1 mL of serum to 20 mL with water.
2. For estimating **total bilirubin**, mark two tubes as test (Tt) and blank (Tb).
3. For **direct bilirubin**, label two test tubes as test (Td) and blank (Db).
4. For **standard**, label two test tubes as standard (S) and blank (Sb). The procedure is summarised in the table below.
5. Mix well and keep for 30 minutes. Then read the OD at 540 nm.

Reagents (mL)	Tt	Tb	Td	Db	S	Sb
Methanol	5	5	–	–	8.8	8.8
Water	–	–	5	5	–	–
Diazo reagent	1	–	1	–	1	–
1.5% HCl	–	1	–	1	–	1
Diluted serum	4	4	4	4	–	–
Bilirubin standard	–	–	–	–	0.2	0.2

Calculation: Total bilirubin (mg/dL)

$$Tt = \frac{\text{OD of Tt} - \text{OD of Tb}}{\text{OD of S} - \text{OD of Sb}} \times 0.2 \times \frac{20}{4} \times 100$$

$$= \frac{\text{OD of Tt} - \text{OD of Tb}}{\text{OD of S} - \text{OD of Sb}} \times 100$$

Direct bilirubin (mg/dL)

$$= \frac{\text{OD of Tt} - \text{OD of Tb}}{\text{OD of S} - \text{OD of Sb}} \times 100$$

Clinical significance

Elevated level of bilirubin in blood is known as **hyperbilirubinemia**. Based on the increased levels of conjugated and unconjugated bilirubin, the conditions can be classified as follows.

1. Increased levels of unconjugated bilirubin (indirect hyperbilirubinemia)
   For example, hemolytic jaundice, Criggler–Najjar syndrome.
2. Increased levels of conjugated bilirubin (direct hyperbilirubinemia). This occurs in obstructive jaundice. The obstruction is of two types.
   (a) Intrahepatic obstruction
      (i) Hepatoma of liver
      (ii) Lymphoma of liver
      (iii) Viral hepatitis
   (b) Extrahepatic obstruction
      (i) Cholestasis (gall bladder stones)
      (ii) Carcinoma of head of pancreas
3. Increased levels of both conjugated and unconjugated bilirubin (biphasic hyperbilirubinemia). Seen in hepatitis.
   (a) Infective hepatitis – occurs due to infection by the hepatitis virus.
   (b) Alcoholic hepatitis – occurs due to regular consumption of large quantities of alcohol.

Viva Questions

1. What is bilirubin and how is it formed in the body?
2. What are direct and indirect bilirubin?
3. What are the normal levels of direct, indirect and total bilirubin in blood?
4. What is the Van den Bergh reaction?
5. Name the enzyme responsible for conjugation of bilirubin.
6. What are the common types of jaundice?
7. What are the causes of hemolytic, hepatic and obstructive jaundice?
8. What is neonatal physiologic jaundice? What is the rationale of use of phototherapy to treat neonatal jaundice?
9. What are the congenital disorders of bilirubin metabolism?
10. Why do patients pass clay-coloured stools in obstructive jaundice?

14 Aminotransferases (AST and ALT or SGOT and SGPT)

BI 11.13

Aminotransferases (earlier called transaminases) are an important group of enzymes of amino acid metabolism that catalyse the transfer of an amino group from an amino acid to a keto acid.

The body contains several aminotransferases of which the following two are of diagnostic importance:

1. Aspartate aminotransferase (AST) or serum glutamate oxaloacetate transaminase (SGOT)
2. Alanine aminotransferase (ALT) or serum glutamate pyruvate transminase (SGPT)

Several methods are available for estimating aminotransferases.

Estimation of AST and ALT by the Reitman and Frankel Method

Principle: To estimate AST, serum is treated with aspartate and α-ketoglutarate, whereas to measure ALT, alanine and α-ketoglutarate are used as substrates. The two new ketoacids formed in the reactions (oxaloacetate and pyruvate, respectively) are treated with 2, 4-dinitrophenylhydrazine. The absorbance of the brown-coloured dinitrophenyl- hydrazones is measured under alkaline conditions at 505 nm. The increase in absorbance at 505 nm is directly proportional to the activity of the enzymes.

Reagents used

1. Phosphate buffer (0.1 M), pH 7.4
2. Substrates
   (a) **For AST**: This substrate contains 2 mM of α-ketoglutarate and 100 mM of L-aspartate per litre. Dissolve 29.2 mg of α-ketoglutarate and 1.33 g of L-aspartate in 1 N NaOH with stirring. Adjust to pH 7.4 with NaOH and make up to 100 mL with phosphate buffer (0.1 M, pH 7.4).
   (ii) **For ALT**: This substrate contains 2 mM of α-ketoglutarate and 100 mM of L-alanine per litre. Dissolve 29.2 mg of α-ketoglutarate and 0.89 g of L-alanine in 1 N NaOH with stirring. As mentioned above, adjust the pH to 7.4 and make up to 100 mL.
3. **Pyruvate standard**: It contains 2 mM of sodium pyruvate per litre. Dissolve 22 mg of sodium pyruvate in 100 mL phosphate buffer. 0.2 mL of this standard used in the estimation is equivalent to 0.4 m moles. Note that for AST assay also pyruvate rather than oxaloacetate is used as the standard, because oxaloacetate is known to be unstable and gets converted to pyruvate during the course of estimation.

4. **Colour reagent (DNPH)**: Dissolve 19.8 mg of 2, 4-dinitrophenylhydrazine in 100 mL of 1 N HCl and store in a brown bottle.
5. NaOH (0.4 N) – see Table 14.1.

Procedure

Constitute four test tubes as blank, standard, test and control as shown below. The control test tube is for comparison.

Table 14.1 *Estimation of AST and ALT by the Reitman and Frankel method*

Reagent (in mL)	Test	Control	Standard	Blank
Buffered substrate	1	1	1	1
Serum	0.2	–	–	–
Standard	–	–	0.2	–
Distilled water	–	–	–	0.2
Incubate at 37°C for 60 minutes for AST and 30 minutes for ALT				
Serum	–	0.2 mL	–	–
DNPH	1	1	1	1
Mix well and allow to stand for 20 minutes at room temperature				
NaOH (0.4 N)	10	10	10	10
After 20 minutes, read at 505 nm.				

Calculation

Concentration of AST/ALT (IU)

$$= \frac{\text{OD of test} - \text{OD of control}}{\text{OD of standard} - \text{OD of blank}} \times 0.4 \times \frac{1000}{0.2 \times \text{time (minutes)}}$$

The concentration is also expressed in karmen units (1 karmen unit = 0.483 I.U).

Spectrophotometric (UV Range) Methods

In these methods of estimating ALT, the pyruvate formed is converted to lactate by lactate dehydrogenase (LDH) in the presence of NADH.

$$\text{Pyruvate} + \text{NADH} \xrightarrow{\text{LDH}} \text{Lactate} + \text{NAD}^+$$

Decrease in absorbance at 340 nm is directly proportional to ALT activity.

In the same fashion, oxaloacetate formed during the action of AST is reduced to malate by malate dehydrogenase (MDH) in the presence of NADH.

$$\text{Oxaloacetate} + \text{NADH} \xrightarrow{\text{MDH}} \text{Malate} + \text{NAD}^+$$

Here, again, decrease in absorbance at 340 nm is proportional to AST activity.

Clinical significance

The heart, liver and skeletal muscles and kidneys are rich in AST whereas ALT is high in the liver. So it could be expected that serum levels of AST will increase due to necrosis and inflammation of the heart, liver and muscle whereas ALT levels can be expected to rise in liver damage.

Normal serum levels of AST – 8–20 IU/L

Normal serum levels of ALT – 15–40 IU/L (in men)
10–30 IU/L (in women)

- The levels of AST are significantly elevated in **myocardial infarction** (3–9 hours after the onset of attack). Its levels are moderately elevated in liver diseases (marked increase is observed in primary hepatoma), muscular dystrophy and acute pancreatitis.
- Both AST and ALT are increased in **liver disease**, but ALT levels are more marked than AST levels. **Hepatocellular jaundice** is characterised by increased levels of both ALT and AST.
- Very high levels of ALT are observed in **acute hepatitis** (both viral and toxic origin). Moderate increase is seen in *cirrhosis and malignancy of the liver.*
- Low values of AST are observed in vitamin B_6 deficiency.

Viva Questions

1. What is transamination? Name two aminotransferases with clinical significance.
2. Name the amino acids which do not undergo transamination.
3. Name the substrates and products of the transamination reactions catalysed by ALT and AST.
4. Name the coenzyme that is essential for transamination reactions.
5. What are the normal levels of ALT and AST? What is their clinical significance?
6. What is the principle of estimation of ALT and AST by the Reitman and Frankel method?

15 Alkaline and Acid Phosphatases (ALP and ACP)

BI 11.14

Phosphatases are enzymes that remove phosphoric acid from a large number of substrates. There are two types of phosphatases:

1. Alkaline phosphatase, which has maximum catalytic activity at alkaline pH (about 10).
2. Acid phosphatase, which exerts maximum activity at acidic pH (between 5 and 6).

Alkaline phosphatase (ALP) is involved in *bone calcification*. It is also associated with *lipid transport* in the intestine. ALP measurements reflect the combined activity of several alkaline phosphatase isoenzymes derived from the liver, bones, kidneys, intestine and placenta. The predominant forms present in normal serum are the liver and bone varieties.

ALP levels in serum are measured to

- detect skeletal disease
- check for liver lesions that cause biliary obstruction
- assess response to vitamin D in the treatment of rickets.

Acid phosphatase (ACP) is found primarily in the prostate gland and semen. It also occurs in low quantities in the liver, spleen, RBCs and bone marrow.

ACP levels in serum are estimated to

- detect prostate cancer.

Estimation of Alkaline Phosphatase

Principle

Disodium phenylphosphate releases phenol by enzymatic hydrolysis under specific conditions of pH, temperature and time. This reacts with 4-aminoantipyrine in the presence of an alkalline oxidising agent, producing a red-coloured compound which is measured at 520 nm. NaOH is added after incubation to raise the pH to stop the reaction. Potassium ferricyanide acts as the oxidising agent. Sodium bicarbonate provides the alkaline medium.

Reagents used

- *Substrate (disodium phenylphosphate)*
    1. Dissolve 2.18 g (2.54 g of the dihydrate) in distilled water and make the volume up to 1 L.
    2. Boil the solution to eliminate microorganisms and add chloroform (4 mL/L).
    3. Keep in a brown bottle at 4°C.

- *Buffer (sodium carbonate and sodium bicarbonate)*
  1. Dissolve 6.3 g of anhydrous sodium carbonate and 3.36 g of sodium bicarbonate per litre in water.
  2. Keep at 4°C. pH of this buffer is 10.
- *0.5 N NaOH:* 20 g of NaOH per litre in distilled water.
- *0.5 N sodium bicarbonate:* 42 g/L in water.
- *Potassium ferricyanide:* 24 g/L in distilled water. Store in a brown bottle.
- *4-Aminoantipyrine:* 6 g of 4-aminoantipyrine per litre in water. Store in a brown bottle.
- *Stock standard (phenol solution 1 g/L)*
  1. Dissolve 1 g of crystalline phenol in 0.1 N HCl and make up to 1 L with acid.
  2. Keep at 40°C in a brown bottle (1N HCl = 90 mL/HCl/L).
- *Phenol working standard (1 mg/100 mL)*
  1. Dilute the stock standard 1 mL to 100 mL with distilled water.
  2. Preserve it by adding a few drops of chloroform and keep at 40°C in a brown bottle.

Procedure

1. Label four test tubes as blank, standard, test and control.
2. Estimate as shown below.

Reagent (in mL)	Blank	Standard	Test	Control
Buffer	1.1	1.1	1	1
Substrate	–	–	1	1
Incubate at 37°C for 3 minutes				
Distilled water	1	–	–	–
Standard	–	1	–	–
Serum	–	–	0.1	–
Incubate at 37°C for 15 minutes				
0.5 N NaOH	0.8	0.8	0.8	0.8
Serum	–	–	–	0.1
Sodium bicarbonate	1.2	1.2	1.2	1.2
Mix all the contents				
4-Aminoantipyrine	1	1	1	1
Potassium ferricyanide	1	1	1	1
Avoid exposure to direct sunlight. Read at 510 nm.				

Calculation

The concentration of alkaline phosphatase is expressed in king Armstrong (KA) units. Each KA unit represents the release of 1 mg of phenol per 100 mL of serum in 60 minutes at 37°C.

Concentration of ALP (KA/100 mL)

$$= \frac{\text{OD of test} - \text{OD of control}}{\text{OD of standard} - \text{OD of blank}} \times \frac{\text{Concentration of phenol in standard}}{\text{Volume of the test}} \times 100$$

$$= \frac{T - C}{S - B} \times \frac{10\ \mu g}{0.1} \times 100$$

$$= \frac{T - C}{S - B} \times 100$$

Clinical significance

The normal range of ALP is 3–13 KAU/dL (40–125 IU/L). The levels of ALP in childhood and adolescence are almost double the normal adult values. The levels of ALP are primarily increased in **bone disease and biliary tract disease.**

The activity of ALP is increased during active bone growth and bone regeneration. The following conditions are characterised by elevated ALP levels: rickets, osteomalacia, Paget's disease and during the healing phase of bone fractures.

The liver synthesises more ALP when there is an obstruction in the biliary tract. *Extrahepatic obstruction by gall stones or malignancy of the head of pancreas* is characterised by marked elevation in ALP levels. *Intrahepatic obstruction* also results in increased levels of ALP.

A decrease in ALP levels is seen in kwashiorkor, severe anemia and vitamin C deficiency (scurvy).

Estimation of Acid Phosphatase

The estimation is almost identical to that of ALP. All the reagents except the buffer used for ALP can be used for ACP.

Citrate buffer for ACP (0.2 mol/L), pH 5.5 at 20°C

1. Dissolve 42 g (0.2 mol) of citrate in 800 mL of distilled water.
2. Add 40 mL of NaOH (7.5 mol/L).
3. Mix well and adjust the pH to 5.5 with acid or alkali as per the requirement.
4. Make the total volume up to 1 L. The buffer can be stored at 4°C for up to six months.

Procedure

1. Label four test tubes as blank, standard, test and control.
2. Estimate as shown below.

Reagent (in mL)	Blank	Standard	Test	Control
Buffer	1.2	1.2	1	1
Substrate	–	–	1	1
Incubate at 37°C for 3 minutes				
Serum	–	–	0.2	–
Distilled water	1	–	–	–
Working standard	–	1	–	–
Incubate for 60 minutes				
NaOH	1	1	1	1
Serum	–	–	–	0.2
$NaHCO_3$	1	1	1	1
4-Aminoantipyrine	1	1	1	1
Potassium ferricyanide	1	1	1	1

Calculation

ACP levels are expressed in KA units. One KA unit corresponds to the production of 1 mg of phenol liberated by 100 mL of serum in 60 minutes at 37°C.

$$\text{Concentration of ACP (KAU/dL)} = \frac{T-C}{S-B} \times \frac{\text{Concentration of standard}}{\text{Volume of the test}} \times 100$$

$$= \frac{T-C}{S-B} \times \frac{100\ \mu g}{0.2} \times 100$$

$$= \frac{T-C}{S-B} \times 5.$$

Clinical application

Normal range of acid phosphatase is 0.5–4 KAU/dL (2.5–12 IU/L). Marked elevation of ACP levels are observed in **prostate cancer.**

Viva Questions

1. What are the different sources of ALP in the body?
2. Why are the levels of ALP elevated in rickets and obstructive jaundice?
3. What is Regan's isoenzyme?
4. What are the normal levels of ACP and in which condition are its levels elevated?

16

BI 2.5; 2.6; 2.7; 11.17

Creatine Kinase and Its Isoenzymes

Creatine kinase (CK) plays a role in energy production during muscle contraction. Its levels reflect normal tissue destruction (turnover). A high CK level in blood is indicative of injury to tissues, as occurs in a heart attack.

Creatine kinase consists of two subunits derived from muscle (M) and brain (B). These combine to form three isoenzymes:

1. CK-MM – Mostly occurs in skeletal muscle
2. CK-MB – Primarily found in heart muscle
3. CK-BB – Found mostly in brain tissue

Isoenzyme measurements are used to determine the precise location of tissue destruction *(acute heart attack is characterised by high levels of CK-MB).*

Indications for CK estimation

- To detect and diagnose **acute myocardial infarction** and recurrent episodes of the same.
- To evaluate the patient for possible causes of chest pain and to confirm or rule out heart attack (MI).
- To detect skeletal muscle disorders and evaluate skeletal muscle injury.

Estimation of Creatine Kinase by Colorimetric Method (Hughes)

Principle

This method measures the colour produced by creatine with diacetyl and alpha-naphthol:

Creatine phosphate + ADP ⟶ Creatine + ATP

Creatine + Diacetyl + Naphthol ⟶ Coloured complex

1. Mg^{2+} is used as an activator.
2. Cysteine is used to reactivate the sulphydryl groups of CK.
3. Barium hydroxide and zinc sulphate are used to precipitate proteins and to prevent acid hydrolysis of creatine phosphate.
4. Chloromercuribenzoate is added to complete the inactivation of CK and to prevent interference of cysteine with subsequent colour development.

Reagents used

1. Creatine phosphate – 12 mmol/L, 28 mg sodium creatine phosphate per 10 mL of the buffer diluted to 10.
2. Magnesium acetate solution – 25 mmol/L, 5.4 g of tetrahydrate salt in 1 L of buffer.

3. Adenosine-5-diphosphate (4 mmol/L) – Dissolve 23 mg of trisodium ADP. $5H_2O$ in 10 mL water.
4. Tris buffer (100 mmol/L, pH 7.4) – Dissolve 12.1 g of Tris in 820 mL mmol/L HCl and make it up to 1 L with water. Adjust pH to 7.4.
5. p-chloromercuribenzoate solution (30 mmol/L)
   (a) Dissolve 1.07 g in 25 mL 1 mol/L of sodium hydroxide.
   (b) Add about 22 mL 1 mol/L of HCl, drop by drop, while shaking until a slight precipitate forms. This dissolves on diluting to 100 mL with water. The pH should be 10.0–10.5.
6. Cysteine (150 mmol/L) – Prepare freshly by dissolving 95 mg of cysteine hydrochloride in 0.6 mL mmol/L sodium hydroxide and add 3.4 mL of buffer.
7. Barium hydroxide (150 mmol/L)
   (a) Boil 6 g $Ba(OH)_2$ in 100 mL of water for a few minutes.
   (b) Cool, filter and dilute 5 mL of the zinc sulphate solution with about 25 mL of water.
   (c) Titrate with barium hydroxide using phenolphthalein as an indicator.
8. α-naphthol solution – 160 mg/10 mL in an alkali solution, prepare freshly.
9. Zinc sulphate solution – 50 g/L of $ZnSO_4$. $7H_2O$.
10. Stock alkali solution – 60 g sodium hydroxide and 128 g anhydrous sodium carbonate in 1 L of water.
11. Stock standard creatine solution (6 mmol/L) – 895 mg creatine hydrate/L in water. Keep in the refrigerator.
12. Working standard creatine solution – 1.2 mmol/L – freshly prepared by diluting the stock standard 1 to 5.
13. Stock diacetyl solution – 1 mL/100 mL of water. Keep in a brown bottle.
14. Working diacetyl solution – dilute the stock solution 1 to 20 with water.

Procedure

1. Take four tubes labelled test (T), control (C), standard (S) and blank (B).
2. Perform the following operations as shown below.

Reagent (in mL)	T	C	S	B
Creatine phosphate	0.25	0.25	–	–
Magnesium solution	0.20	0.20	0.20	0.20
Cysteine	0.10	0.10	0.10	0.10
Serum	0.10	0.10	–	–
Mix. Preincubate at 37°C for a few minutes				
ADP	0.25	–	–	–
Mix. Incubate for 30 minutes. In the meantime, add:				
ADP	–	0.25	–	–
Chloromercuribenzoate	–	0.50	–	–
Creatine std, 1.2 mmol/L	–	–	0.25	–
Water	–	–	0.35	0.60

Reagent (in mL)	T	C	S	B
After exactly 30 minutes stop the reaction in tube 'T'				
Chloromercuribenzoate	0.50	–	0.50	0.50
Zinc sulphate	0.50	0.50	0.50	0.50
Barium hydroxide	0.50	0.50	0.50	0.50
Centrifuge at 5,000 rpm for 15 minutes				
Supernatant	0.10	0.10	0.10	0.10
α-naphthol	2.50	2.50	2.50	2.5
Diacetyl	0.50	0.50	0.50	0.50
Water	6.00	6.00	6.00	6.00

Incubate at 37°C for 1 hour. Mix by inversion every 15 minutes. Read at 520 nm. Record the absorbances of the 4 tubes.

Calculations

A unit of creatine kinase activity (U) is defined as that which catalyses the conversion of 1.0 μmol of substrate (creatine phosphate) to product (creatine) in 1 minute under standard conditions of pH and temperature.

Amount of creatine formed in tube 'T',

$$\text{Creatine} = \frac{T - C}{S - B} \times 1.2 \times 0.25\ \mu\text{mol}$$

$$\text{Creatine kinase activity (U/L)} = \frac{\text{Creatine} \times 1000}{30 \times 0.1}$$

Clinical application

Normal range of creatine kinase in serum:

15–100 U/L (male)
10–80 U/L (female)

CK-MB levels rise after **myocardial infarction,** reach their peak in 12–24 hours and return to normal in 24–48 hours. CK-MM levels elevate following skeletal muscle injury. In acute psychotic episodes, CK-MM levels shoot up due to increased muscular activity caused by agitation. CK-BB levels are increased in brain tissue injury, severe shock and malignant tumours. Total CK levels are elevated in hypokalemia, alcoholic cardiomyopathy and carbon monoxide poisoning.

Viva Questions

1. What are creatine, creatinine and phosphocreatine?
2. What is the role of CK in the body? What is *Lohmann's reaction*?
3. What are the normal levels of CK? In which conditions are CK levels in blood elevated?
4. What are the isoenzymes of CK?
5. What is the reaction mechanism of estimation of CK by the Hughes method?

17

BI 2.5; 2.6; 2.7; 11.17

Amylase and Lipase

The enzyme amylase plays a role in the digestion of starch and glycogen in the mouth, stomach and intestine. In suspected pancreatic disease, measurement of amylase in blood and urine plays an important role in diagnosis.

Lipase is the enzyme produced in the pancreas and is secreted into the duodenum where it converts triacylglycerols and other dietary lipids into free fatty acids and glycerol. Destruction of pancreatic tissue, which occurs in **acute pancreatitis,** releases large amounts of amylase and lipase into blood.

Indications for the estimation of serum amylase

- To diagnose **acute pancreatitis**
- To distinguish between acute pancreatitis and other causes of acute abdomen that require immediate surgery
- To detect possible pancreatic injury caused by abdominal trauma in accidents

Indications for the measurement of serum lipase

- To aid in the diagnosis of acute pancreatitis

Estimation of Serum Amylase

Principle

Starch reacts with iodine to give a blue-coloured complex. Amylase acts upon starch to produce maltose along with some maltotriose and glucose. The difference in intensity of the blue colour produced before and after adding iodine solution to a fixed quantity of serum–starch substrate mixture is directly proportional to the serum amylase level.

Reagents used

1. *Buffered starch substrate (pH 7.0)*
   (a) Dissolve 13 g of dry anhydrous disodium hydrogen phosphate (or 33.5 g $Na_2HPO_4/12H_2O$ and 4.3 g of benzoic acid) in 250 mL of water. Boil the solution.
   (b) Mix 0.2 g of soluble starch in 5–10 mL of cold water in a beaker.
   (c) Add this starch solution to the boiling mixture with additional cold water.
   (d) Continue to boil for one minute. Cool to room temperature and dilute to 500 mL.
   (e) Store at 4°C. The solution is stable for one month.
2. *Stock iodine solution (0.1 N)*
   (a) Dissolve 13.5 g of sublimed iodine in a solution of 24 g of potassium iodide in about 100 mL of water.
   (b) Dilute to 1 L with water.

3. *Working iodine solution (0.01 N)*
   (a) Dissolve 50 g of potassium fluoride in a little water.
   (b) Add 100 mL of stock iodine solution and make it up 1 L with water.
   (c) Store at 40°C in a brown bottle.

Procedure

1. Dilute the serum 1 in 10 with 0.9% saline.
2. *Test*: Take 1 mL of buffered starch substrate in a test tube. Keep it in a water bath at 37°C for 4 minutes. Add 0.1 mL of diluted serum and mix gently. Incubate for 15 minutes at 37°C. Remove the tube from the bath. Then add 0.4 mL of working iodine solution. Mix well and add 8.5 mL of distilled water. Mix thoroughly.
3. *Control*: Mix together in order 1 mL of buffered substrate, 8.6 mL of distilled water and 0.4 mL of iodine in a test tube marked control (C). Measure the optical density of test and control at 660 nm.
4. Use water as *blank*.

Calculation

One Somogyi unit of amylase activity is defined as the amount of enzyme digested in 5 mg of starch in the abovementioned conditions. The control tube contains 0.4 mg of starch. The amount of starch which has been digested can be calculated as follows:

$$\frac{C-T}{C-B} \times 0.4 \text{ mg}$$

The quantity of enzyme present in 0.01 mL of serum is:

$$\frac{C-T}{C-B} \times \frac{0.4 \text{ mg}}{5} \text{ Somogyi units.}$$

Thus, quantity of serum amylase is:

$$\frac{C-T}{C-B} \times \frac{0.4}{5} \times \frac{100}{0.01}$$

$$\frac{C-T}{C-B} \times 800 \text{ Somogyi units.}$$

Urinary amylase: The serum amylase method is applicable to urine. Dilution of a urine sample with sodium chloride in the ratio 1:1 or 1:2 is done before subjecting the urine sample to the analytical process.

Clinical significance

The normal serum amylase level is up to 200 Somogyi units/100 mL. Increase in serum amylase occurs in acute pancreatitis, reaching levels up to 2000 Somogyi units/100 mL or even more. Levels exceeding 550 Somogyi units/100 mL support a diagnosis of **acute pancreatitis.** The peak serum level is seen within 12–24 hours after the onset of an acute pancreatitic attack. The values return to normal in 4–8 days. Serum amylase levels that remain high or increase again later suggest the formation of a pseudocyst of the pancreas.

Serum amylase is also raised in parotitis, chronic cholecystitis, hepatitis, cirrhosis and ruptured ectopic pregnancy. The level of enzyme in all these cases is moderately increased and is usually less than three times for the upper normal limit.

Estimation of Serum Lipase

Principle

Lipase acts on a known amount of standard olive oil or triolein to liberate oleic acid. The liberated oleic acid is estimated by titration using thymolphthalein as the indicator.

Reagents used

Substrate emulsion

1. Mix 100 mL of water with 0.2 g of sodium benzoate and 7 g of gum acacia in a mechanical blender at minimum speed until it is dissolved.
2. Add 100 mL of purified olive oil and mix for 10 minutes at maximum speed.
3. Store in a refrigerator at 4°C. The emulsion may be kept for up to one month.
    - Tris buffer: 200 mmol/L (pH 7.8 at 37°C)
    - Sodium hydroxide: 50 mmol/L
    - Aqueous ethanol: 950 mL ethanol/L
    - Thymolphthalein solution: 1 g in 100 mL aqueous ethanol

Procedure

1. Take two test tubes and mark them 'test' (T) and 'control' (C), respectively.
2. Add 2.5 mL of water, 10 mL of substrate emulsion and 1.0 mL of Tris buffer.
3. Keep both tubes in a water bath at 37°C.
4. Add 1 mL of serum to test. Cover with paraffin and shake vigorously.
5. Place the tubes again in the water bath for three hours. Remove both tubes to stop the reaction and bring the fatty acid into the solution.
6. Then add 5 drops of thymolphthalein solution to each tube and titrate with NaOH to a blue colour (greenish blue with icteric serum). Better results are obtained if the titration is done using a pH meter to an end point of pH 10.5 instead of using thymolphthalein as indicator.

Calculation

Serum lipase (Cherry–Crandall units/mL) = Titration of test (mL) – Titration of control (mL). Under the above mentioned conditions, 1 Cherry–Crandall unit/mL is equivalent to 227/U/L.

Lipase unit: 1 Cherry–Crandall unit is the enzymic activity which will liberate fatty acids equivalent to 1 mL of 50 mmol/L sodium hydroxide.

Clinical significance

Normal serum lipase level is 0.01–1.0 CCU/mL. In **acute pancreatitis,** serum lipase is elevated and the increase may be up to 12 CCU/mL using 24 hours' incubation. Changes in serum lipase and amylase are roughly parallel in patients with acute pancreatitis. However, serum lipase remains raised for longer than serum amylase as the former is removed from circulation more slowly than the latter. Therefore, serum lipase estimation is helpful in diagnosing acute pancreatitis several days after the onset of attack.

Serum lipase is elevated in acute intra-abdominal conditions such as acute appendicitis and paralytic ileus and also in renal failure. Lipase levels are often significantly increased in fractures of the bones and in fat embolism.

Viva Questions

1. What are the indications for the estimation of serum amylase and serum lipase?
2. What is the normal level of serum amylase?
3. What is the principle of estimation of serum amylase?
4. How is serum amylase estimation related to acute pancreatitis?
5. What is the normal serum level of lipase? What is its clinical significance?

18 Calcium and Phosphorus

BI 11.11

Calcium performs many functions in the body. It helps regulate nerve, muscle and enzyme activity. More than 98% of the body's calcium is found in the bones and teeth. Calcium homeostasis is one of the finely regulated mechanisms in the human body.

Phosphorus is essential for bone formation along with calcium. It plays a role in various metabolic pathways, regulating acid–base balance. It is an integral component of the energy currency of the cell, ATP.

Calcium and phosphorus levels are measured for the diagnosis of various skeletal, neuromuscular and endocrine disorders, acid–base imbalances, blood-clotting deficiencies, irregular heart rhythms and for kidney disorders.

Estimation of Serum Calcium

Principle

Ca^{2+} forms a blue-coloured complex with Arsenazo-III. The intensity of the colour is directly proportional to the concentration of the Ca^{2+} present in the given sample. The intensity is measured at 650 nm.

Reagents used

1. Arsenazo-III reagent
2. 8-hydroxy quinolone reagent
3. Ca^{2+} standard – 8 mg/dL

Procedure

Constitute blank, standard and test as shown below.

	Blank	Standard	Test
Reagent	1.0 mL	1.0 mL	1.0 mL
Distilled water	0.02 mL	–	–
Standard	–	0.02 mL	–
Sample	–	–	0.02 mL

Mix well and keep at room temperature for 2 minutes. Take OD readings at 650 nm.

Calculation

$$\text{Concentration of the } Ca^{2+} \text{ present in the given sample (mg/dL)} = \frac{C - T}{C - B} \times 8.$$

Normal value: 8.5–11.0 mg/dL

Clinical significance

Elevated (↑) *Ca levels are seen in*

1. Hyperparathyroidism
2. Metastatic carcinoma of the bone (Paget's disease)
3. Multiple myeloma
4. Hypervitaminosis – D
5. Multiple fractures
6. Prolonged immobilisation

Lowered (↓) *Ca levels are seen in*

1. Tetany
2. Rickets
3. Kidney failure
4. Malabsorption syndromes
5. Medullary carcinoma of the thyroid
6. Acute pancreatitis and peritonitis

Estimation of Serum Phosphorus

Principle

Ammonium molybdate under acidic conditions reacts with phosphorus to form a phosphomolybdate complex which is reduced to a blue colour by a metal. The intensity of the blue is directly proportional to the concentration of phosphorus present in the given serum sample. The intensity is measured at 680 nm.

Reagents used

1. Buffer/catalyst
2. Moloybdate reagent
3. Metal reagent
4. Phosphorus standard–5 mg/dL

Procedure

Constitute blank, standard and test as shown below.

	Blank	Standard	Test
Buffer	1.0 mL	1.0 mL	1.0 mL
Molybdate reagent	1.0 mL	1.0 mL	1.0 mL
Distilled water	0.1 mL	–	–
Standard	–	0.1 mL	–
Sample	–	–	0.1 mL
Metal reagent	1.0 mL	1.0 mL	1.0 mL

Mix and keep at room temperature for 5 minutes. Take the OD reading.

Calculation

Concentration of phosphorus present in the given serum sample (mg/dL)

$$= \frac{C - T}{C - B} \times 5$$

Normal value: 2.5–5.0 mg/dL.

Clinical significance

Elevated (↑) *phosophorus level is seen in*

1. Hyperparathyroidism
2. Rickets
3. Osteomalacia
4. Kidney tubular acidosis

Lowered (↓) *phosphorus levels are seen in*

1. Chronic nephritis
2. Hypoparathyroidism
3. Skeletal disease
4. Kidney failure

Serum calcium and phosphate levels have an inverse relationship—if one is elevated, the other is lowered. Phosphate levels alone are of limited diagnostic value, so they are always interpreted with calcium levels.

Viva Questions

1. What are the biochemical functions of calcium in the body?
2. What are the hormones that regulate blood calcium levels?
3. What are the dietary sources and RDA of calcium?
4. What are the causes of hypercalcemia and hypocalcemia?
5. What is *tetany*?
6. What precautions do we need to take while injecting calcium through the IV route? Why do we need to take these precautions?
7. What is osteomalacia and osteoporosis?
8. What is the biochemical role of phosphate in the body?
9. What are the dietary sources of phosphate?
10. What are high-energy phosphates?
11. Name the scientist who proposed the concept of high-energy compounds.
12. Describe the principle of estimation of serum calcium.
13. Describe the reaction mechanism of estimation of serum calcium.

19 Serum Electrolytes

BI 11.16

Body fluids including blood contain many cations (Na^+, K^+, Ca^{2+}) and anions (Cl^-, PO_4^{3-}, HCO_3^-). Among these, Na^+, K^+ and Cl^- in serum/plasma are commonly measured to evaluate different clinical conditions. These are known as serum electrolytes. Electrolytes are given orally and as intravenous fluids on many occasions to maintain electrolyte balance and to detect abnormalities associated with electrolyte balance.

The flame photometer is used to estimate the sodium and potassium ions in body fluids. Today, *electrolyte analysers* with ion-selective electrodes are being increasingly used for determining sodium and potassium levels.

Estimation of sodium and potassium by flame photometer principle by ISE method

When a metal is heated, due to heat energy, the electrons in the outer orbitals of atoms absorb energy and move to a higher energy state. These excited electrons, while returning to the ground state, emit the absorbed energy as radiation of a particular wavelength. The amount of light emitted is directly proportional to the number of excited atoms, which in turn is proportional to the concentration of the substance in the sample.

A flame photometer contains an atomiser or nebuliser, burner, monochromator, photodetector and an output device (Fig. 19.1).

Reagents used

1. Sodium stock standard (200 m Eq/L)
2. Potassium stock standard (10 m Eq/L)
3. Working standards of sodium and potassium

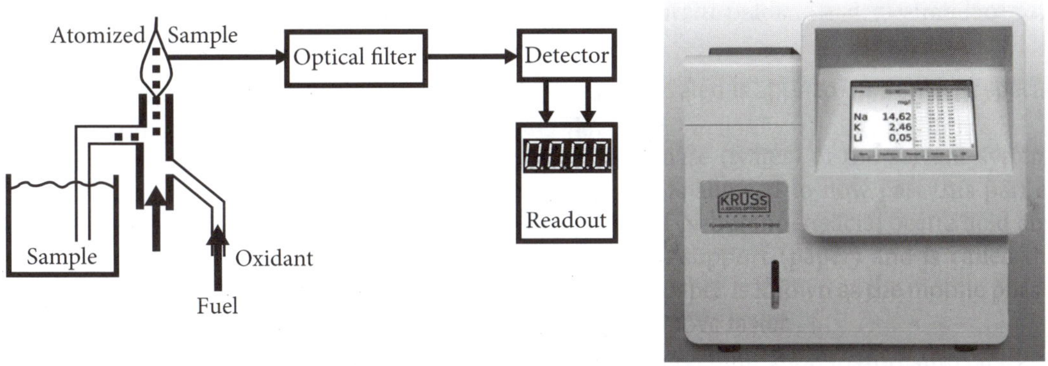

Fig. 19.1 *Flame photometer*

(Image: A.KRÜSS Optronic GmbH / CC BY-SA 3.0 DE; https://creativecommons.org/licenses/by-sa/3.0/de/deed.en)

Take the stock standards of Na^+ and K^+ and make up the final volume to 100 mL with distilled water to obtain different concentrations as shown below.

Working standard	Stock standard of Na^+ (200 m Eq/L) (mL)	Stock standard of K^+ (10 mEq/L) (mL)	Corresponding Na^+ and K^+ concentration in serum (mEq/L)
1	0.55	0.2	110 and 2
2	0.60	0.3	120 and 3
3	0.65	0.4	130 and 4
4	0.70	0.5	140 and 5
5	0.75	0.6	150 and 6
6	0.80	0.7	160 and 7
7	0.85	0.8	170 and 8

Procedure

1. Dilute the plasma/serum sample with distilled water (1:100).
2. Switch on the flame photometer. Ignite the flame and keep distilled water in a cup under the sample inlet.
3. Adjust the air pressure to get a fine spray of the sample. Adjust the galvanometer reading to 0. Use the appropriate standard and diluted sample and take the galvanometer reading. The flame photometer is calibrated using the above standards, and concentrations of sodium and potassium in the serum/plasma sample are obtained directly on the display.

Electrolyte Analysis by Ion Selective Electrode

Earlier, flame emission spectrophotometry was used for electrolyte analysis but it was cumbersome owing to a large number of maintenance and safety procedures. Advances in electrochemistry has enabled the replacement of this technique with ion selective electrodes (ISEs).

- Electrolytes may be classified as anions which are negatively charged ions that move toward an anode, or cations which are positively charged ions that move toward a cathode.
- Important physiological electrolytes include Na^+, K^+, Ca^{2+}, Mg^{2+}, Cl^-, HCO_3^-, HPO_4^{2-}, SO_4^{2-} and some organic anions, such as lactate.
- Amino acids and proteins in solution also carry an electrical charge; they are usually considered separately from electrolytes.
- Major electrolytes (Na^+, K^+, Cl^-, HCO_3^-) occur primarily as free ions, whereas significant amounts (>40%) of Ca^{2+}, Mg^{2+} and trace elements are bound by proteins, mainly albumin.
- Determination of body fluid concentrations of the four major electrolytes (Na^+, K^+, Cl^-, and HCO_3^-) is commonly referred to as an electrolyte profile.
- ISE is used in clinical chemistry to measure concentration of ions in blood, serum, plasma, CSF and urine.

Functions of Electrolytes

- They affect metabolic processes.
- They maintain osmotic pressure, body pH and hydration of fluid compartments.
- They are involved in the regulation of heart and muscle functions.

Ion Selective Electrode

Principle

An ion-selective electrode (ISE), also known as a specific ion electrode (SIE), is a transducer (or sensor) that converts the activity of a specific ion dissolved in a solution into an electrical potential (Fig. 19.2). The voltage is theoretically dependent on the logarithm of the ionic activity according to the Nernst equation.

Parts of an electrolyte analyser

- Removable reagent pack
- Display screen and keyboard
- Sample probe
- Sensor or electrodes such as sodium sensor, potassium sensor, chloride sensor, space sensor and reference sensor
- Peristaltic pump and tubing

The reagent pack and sensors/ electrodes are replaceable components.

Advantages

- Linear response
- Non-destructive: no consumption of analyte
- Non-contaminating
- Short response time in seconds or minutes. Useful in industrial applications
- Unaffected by colour or turbidity

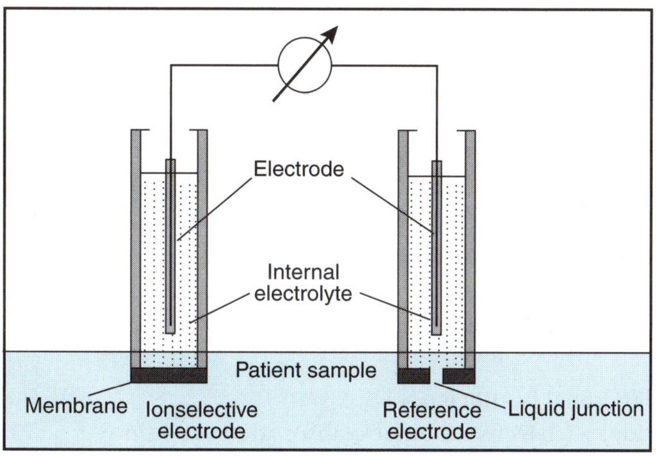

Fig. 19.2 *Setup of an ion selective electrode to measure ionic activity in a solution*

Limitations

- Precision is rarely better than 1%.
- Electrodes can be fouled by proteins and other organic solutes.
- Interference by other ions.
- Electrodes are fragile and have limited shelf life.

Specimens for Electrolyte Determination

- Serum and plasma, obtained from blood collected by venipuncture into an evacuated tube, are the usual specimens analysed for electrolytes.
- Capillary blood, collected in microsample tubes or capillary tubes, or applied directly from a fingerstick to some point-of-care devices, is another sample commonly analysed.
- Heparinized whole blood arterial or venous specimens obtained for blood gas and pH determinations may also be used with direct ion-selective electrodes.
- Urine collection for Na^+, K^+, or Cl^- assays should be done without the addition of preservatives.
- Body fluid aspirates, feces and GI fluid samples may also be submitted for electrolyte analysis.

The following electrolytes are analysed by ISE.

- Sodium (normal value: 136 to 145 mEq/L)
- Potassium (normal value: 3.5 to 5.5 mEq/L)
- Chloride (normal value: 96 to 106 mEq/L)
- Lithium (generally used for diagnosing mental illnesses, including bipolar disorder, depression, and schizophrenia)

Sodium

Sodium is the major cation of extracellular fluid. Na^+ is responsible for almost one half the osmotic strength of plasma. Therefore, it plays a central role in maintaining the normal distribution of water and osmotic pressure in the extracellular fluid compartment (ECF).

The following are the causes of increased and decreased conditions of sodium.

Hypernatremia (increased sodium in blood)

- Excessive intake of salt
- Cushing disease
- Dehydration
- Pregnancy
- Elderly patients with poor intake of water

Hyponatremia (decreased sodium levels in blood)

- Vomiting
- Diarrhea
- Burns
- Addison disease (adrenal insufficiency)
- Renal tubular acidosis (defective tubular reabsorption of sodium)
- Congestive cardiac failure
- Hyperglycemia and ketoacidosis

Potassium

Potassium is the major cation of intracellular fluid. The depolarisation and contraction of heart requires potassium. The following are the causes of increased and decreased conditions of potassium.

Hyperkalemia (increased K^+ level in blood)

- Obstruction of urinary tract
- Renal failure
- Increased hemolysis
- Severe dehydration
- Addision disease
- Metabolic acidosis (diabetic ketoacidosis)
- Drugs like ACE inhibitors, beta blockers

Hypokalemia (decreased K^+ level in blood)

- Cushing syndrome
- Hyperaldosteronism
- Alkalosis
- Insulin therapy
- Diarrhea
- Vomiting
- Intravenous saline infusion in excess
- Drugs like insulin, osmotic diuretics

Chloride

The intake, output and metabolism of sodium and chloride run in parallel. The following are the causes of increased and decreased conditions of chloride.

Hyperchloremia (increased Cl^- level in blood)

- Dehydration
- Cushing syndrome
- Mineralocorticoids cause increased reabsorption from kidney tubules
- Severe diarrhea
- Renal tubular acidosis

Hypochloremia (decreased Cl^- level in blood)

- Excessive vomiting
- Excessive sweating
- In Addison's disease

Viva Questions

1. What are the normal levels of serum sodium, potassium and chloride?
2. Why are K^+ ion abnormalities considered more dangerous than Na^+ abnormalities?
3. What are the common causes of hyponatremia and hypernatremia?
4. Name the common causes of hypokalemia and hyperkalemia.

20 Quantitative Analysis of CSF

BI 11.15

Cerebrospinal fluid (CSF) is found in the sub-arachanoid space, ventricles of the brain and around the spinal cord. The total volume of CSF is about 125 mL in a healthy adult and is renewed every 3–4 hours.

1. Colour: Normal CSF is clear/colourless and gives no sediment on standing.
2. pH: 7.4–7.6

Composition of CSF

Normal constituents

1. Proteins
2. Sugars
3. Chlorides
4. Ca^{+2}
5. Urea
6. Inorganic phosphates
7. Trace amounts of cholesterol

Abnormal constituents

1. Red colour – due to the presence of **RBCs**
2. Yellow colour – due to the presence of **bilirubin**
3. Turbidity – due to the presence of **bacteria,** mainly cocci and infections

Sample collection

CSF is obtained by passing a sterile needle between the L_3 and L_4 vertebrae (**lumbar puncture**).

Determination of Total Proteins in CSF

Principle

Turbidometric methods are most commonly used for the estimation of total proteins in CSF. The proteins present in CSF are precipitated by sulphosalicylic acid. Compare the resulting turbidity with the standard. The turbidity is measured at 660 nm.

Reagents used

1. Sulphosalicylic acid
2. Working standard

Procedure

Take blank, standard and test in the volumes shown below.

	Blank	Standard	Test
Distilled water	0.5 mL	–	–
Standard	–	0.5 mL	–
Sample	–	–	0.5 mL
Sulphosalicylic acid	3.5 mL	3.5 mL	3. 5 mL

Mix well and keep at room temperature for 10 minutes. Measure OD at 660 nm.

Calculation

Concentration of protein in the given CSF sample (mg/dL)

$$= \frac{\text{ODT} - \text{ODB}}{\text{ODS} - \text{ODB}} \times 100.$$

Normal value: 15–45 mg/dL

Clinical significance

Increase in total proteins is the most common abnormality resulting from break-down of blood–CSF and brain– CSF barriers, usually due to inflammatory reactions. The values rise as high as 400 mg/dL in **acute meningitis, polyneuritis, tumours such as neuroma and meningoma.**

Proteins in CSF increase in

- Subarachnoid hemorrhage
- Acute bacterial meningitis
- Tuberculous megningitis
- Viral meningitis } Normal/Elevated
- Multiple sclerosis }

Estimation of CSF Globulins (Pandy's Test)

Principle

The CSF globulins are precipitated by saturated phenol solution (Pandy's reagent). The degree of turbidity is taken as a measure of the concentration of globulins in CSF. Usually it is a method of the visual comparison.

Pandy's reagent: Dissolve 10 g of phenol in 150 mL of distilled water. The reagent should be clear and colourless.

Procedure

1. Add 2 drops of CSF to 2 mL of Pandy's reagent.
2. Observe for turbidity.

Clinical significance

Pandy's test gives a rough idea of the increase in globulins in CSF. Normal fluid shows no turbidity/slight turbidity. But, in conditions like paralysis and multiple sclerosis, where the globulin levels increase to the highest extent, there is marked turbidity.

Estimation of CSF Glucose by Glucose Oxidase and Peroxidase Method

Principle

Glucose is oxidised to gluconic acid and H_2O_2 in the presence of the glucose oxidase enzyme. H_2O_2 splits into H_2O and (O). Nascent O_2 combines with 4-aminoantipyrine to form a red quinone dye in the presence of the peroxidase enzyme. The intensity of the red colour is directly proportional to the concentration of glucose in the given sample. The intensity of the red colour is measured at 540 nm.

Procedure

Take blank, standard and test in the volumes shown below.

	Blank	Standard	Test
Distilled water	0.05 mL	–	–
Standard	–	0.05 mL	–
Sample	–	–	0.05 mL
GOD and POD reagent	5.0 mL	5.0 mL	5.0 mL

Mix well and keep in an incubator at 37°C for 15 minutes.

Calculation

Concentration of glucose in the given CSF sample (mg/dL)

$$= \frac{\text{ODT} - \text{ODB}}{\text{ODS} - \text{ODB}} \times 100.$$

Normal value: 50–80 mg/dL (>60% of blood glucose)

Clinical significance

Lowered (↓) *CSF glucose value is seen in*

1. Bacterial meningitis
2. TB meningitis
3. Malignant meningitis
4. Fungal meningitis
5. Hypoglycemia

Elevated (↑) *CSF glucose is seen in*

1. Encephalitis
2. Poliomyelitis
3. Cerebral abscess
4. Hyperglycemia

Glucose levels in CSF are unaffected in subarachnoid hemorrhage, viral meningitis and multiple sclerosis.

Estimation of Chlorides in CSF

Principle

CSF is titrated with $AgNO_3$ solution using potassium chromate as indicator. When all chlorides are precipitated as AgCl, the excess Ag^+ ions react with the indicator to form a faint brick red colour of silver chromate which is the end point of titration.

Procedure

1. Pipette 1 mL of CSF into a conical flask and add 3 mL of distilled water.
2. Add 2 drops of potassium chromate and mix well.
3. Titrate this with $AgNO_3$ till a faint brick red colour appears.

Calculation

Chlorides present in the given CSF sample (mg/dL) = mL of $AgNO_3$ required for titration × 200.

Normal value: 700–760 mg/dL

Clinical significance

Lowered (↓) *CSF chlorides are seen in*

1. Bacterial meningitis
2. TB meningitis

Elevated (↑) *CSF chlorides are seen in*

1. Hypotension

Viva Questions

1. What is CSF? What is the significance of CSF analysis?
2. How is CSF collected? What is its normal composition?
3. What is the difference between CSF and plasma as regards protein composition?
4. What are the normal levels of proteins, glucose and chlorides in CSF?
5. In which conditions are protein levels in CSF elevated?
6. How are CSF globulins measured?
7. What is the importance of CSF glucose estimation?
8. What is the clinical significance of CSF chloride?

21 Electrophoresis

BI 11.16

Electrophoresis is defined as the migration of charged particles under the influence of an electrical field. A charged particle placed in an electrical field migrates towards one of the electrodes. This property is used for the separation of closely related compounds.

Paper Electrophoresis Technique

- The proteins in serum are separated on the basis of their differential electrophoretic migration due to the difference in their molecular size and charge.
- All the serum proteins carry negative charge at alkaline pH (8.6), and therefore move towards the anode.
- Albumin, being the smallest and carrying maximum negative charge, travels ahead of the globulins.
- α-globulins, in spite of having negative charge, remain at the origin due to their large size.
- β-globulins migrate between albumin γ-globulin.

Apparatus

The horizontal type of electrophoretic tank is divided into an inner buffer compartment and an outer electrode compartment. The platinum electrodes run the full length of the compartment. There is a bridge on which the slide can be placed and wicks can be kept to connect the buffer. In the tank, the buffer level should be same in all the compartments. (Fig. 21.1).

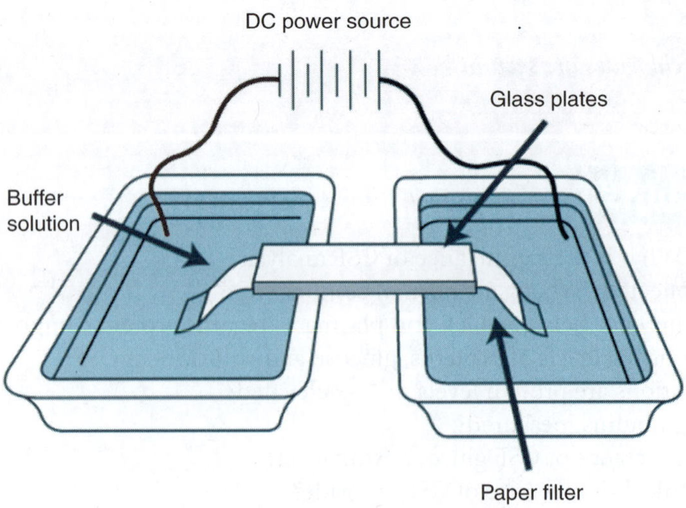

Fig. 21.1 *Apparatus for paper electrophoresis*

Chemicals used

1. Bromophenol blue
2. Ethanol
3. Agarose

Procedure

1. Deliver from a broad-tipped pipette 1.3 mL of warm agar solution on a clean glass slide.
2. Allow the gel to set uniformly and leave at room temperature.
3. Fill the two buffer compartments with equal volumes of 0.05 M barbitone buffer (pH = 8.6). The level should be same in the two compartments.
4. Keep the agar plate in position inside the chamber.
5. Place two filter paper strips of the same width as the agar plate on both ends of the plate and connect to the two buffer compartments.
6. Soak a thin strip of filter paper with the serum sample and place it carefully on the gel at one-third distance from the cathode end of the plate.
7. Close the chamber and switch on the current. Adjust the flow to 3 mA.
8. At the end of two hours, turn off the current and remove the agar plates. Immerse in absolute alcohol for 30 minutes to fix the proteins.
9. Dry the plate and immerse in the bromophenol dye for 2 minutes.
10. Wash at least twice with the wash solution to clear the background colour.
11. Dry the slide and observe the various bands. Quantitation of the electrophoretic result can be done using a **densitometer**.

Applications and clinical significance of electrophoresis

1. Separation of serum proteins
2. Separation of lipoproteins
3. Separation of hemoglobin
4. Separation of isoenzymes
5. For diagnosis of various diseases (Fig. 21.2)

Abnormal Electrophoretic Patterns

Electrophoretic separation of serum proteins indicates various clinical conditions (Fig. 21.2).

1. **Acute infections**: α_1, α_2 fractions of globulins are increased.
2. **Chronic infections**: γ globulins are increased, showing a wide band.
3. **Nephrotic syndrome**: All plasma proteins except α_2 globulin (because of its big size, it is not excreted in urine) are lost in urine and hence a prominent α_2 fraction and diminished albumin are seen.
4. **Multiple myeloma**: A sharp spike termed the **M band (monoclonal band)** is seen in multiple myeloma, a plasma cell cancer. Immunoglobulins (γ globulins) of the same type are produced in large amounts due to the malignant transformation of Ig-secreting cells in this condition. If plasma is used instead of serum, fibrinogen produces a prominent band in the γ region, which may be confused with the M band. In **primary immune deficiency**, the γ-globulin fraction is reduced.

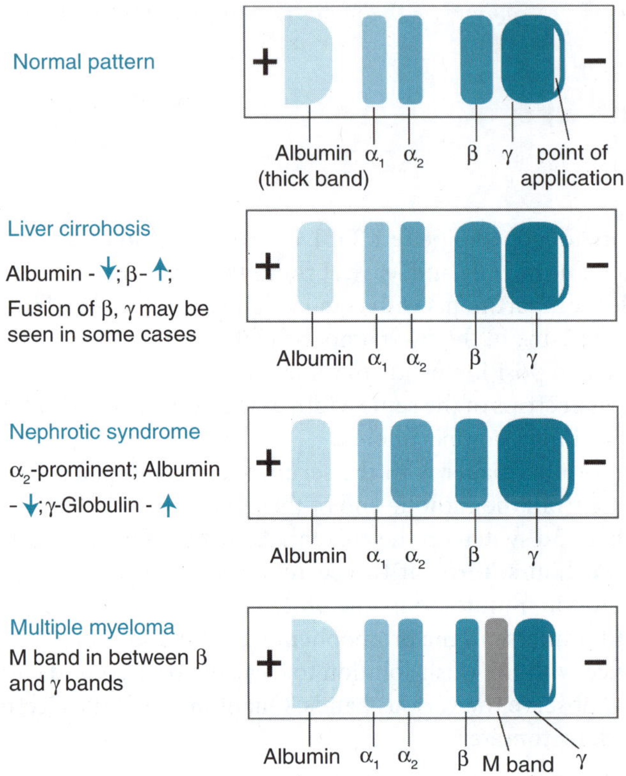

Fig. 21.2 *Electrophoretic separation of plasma proteins in normal individuals and in different pathological states*

5. **Cirrhosis of the liver**: Albumin synthesis by the liver is impaired and a compensatory increased synthesis of globulins by the reticulo-endothelial system takes place. As a result, the albumin band will become thin and a broad β fraction is seen.
6. **α_1-antitrypsin deficiency**: The α_1 band is decreased or sometimes completely disappears.

Types of Electrophoresis

Paper, capillary, gel, discontinuous or disc, isoelectric focusing, and isotachophoresis are the main types of electrophoresis.

1. Paper Electrophoresis

In paper electrophoresis, the sample is applied to a point on a strip of filter paper or cellulose acetate moistened with buffer solution. The ends of the strip are immersed in separate reservoirs of buffer in which the electrodes are placed. Upon application of direct current, the ions of the sample migrate towards the electrodes of opposite polarity to form discrete bands. An ion's migration rate is influenced to some extent by its interaction with the support matrix, but it is largely a function of its charge. Upon completion of the electrophoretogram, the strip is dried and stained to detect the sample components.

In *high voltage paper electrophoresis*, the paper is clamped between two cooled plates to carry off the heat generated by the high-voltage current flow.

Cellulose Acetate Electrophoresis (CAE)

Cellulose acetate paper is the acetate ester of cellulose. The cellulose acetate membrane electrophoresis is most widely used because of its simplicity and the advantage the membrane has over both filter paper and agar. When cellulose reacts with acetic anhydride, it is acetylated to form the raw material for cellulose acetate membranes. The characteristics of the membrane vary with the extent of acetylation. The material is microporous, with pore size ranging from 1 to 5 µm for dry membranes. Such pore sizes prevent spreading of the bands or spots so that plasma proteins are finely resolved even at low voltages. Because of the higher resolution, cellulose acetate is used to separate isoenzymes, lipoproteins, hemoglobins. It allows smaller samples to be examined with high resolution, and its greater sensitivity makes the procedure suitable for screening and visual comparison of many samples applied to a single strip and run for a short time. Minimal protein adsorption eliminates trailing and gives an almost white background after washing. Another advantage of CAE is the speed of separation (20 minutes to 1 hour) and the ability to store the transparent membranes for long periods.

2. Gel Electrophoresis

Gel electrophoresis is among the most powerful and yet conveniently used methods of macromolecular separation. In gel electrophoresis, the molecules to be separated are pushed by an electrical field through a gel that contains small pores. The molecules travel through the pores in the gel at a speed that is inversely related to their lengths. This means that a small DNA molecule will travel a greater distance through the gel than will a larger DNA molecule.

The gels in common use, ployacrylamide, agarose and starch gel, have pores of molecular dimensions whose sizes can be specified. The molecular separation are therefore based on gel filtration as well as the electrophoretic mobilities of the molecules being separated.

Agarose gel electrophoresis (AGE)

This technique is used to analyze serum proteins, hemoglobin variants, lactate dehydrogenase isoenzymes, lipoprotein fractions and other substances. The advantages of agarose gel include its lower affinity for proteins and its native clarity after drying, which permits excellent densitometric examination. Usually, 0.5 to 1.0 g of agarose per deciliter of buffer provides the gel of desired strength with good migration properties. The sample volume required for AGE is relatively less (0.6-3/µL), and electrophoresis time is short (30-90 minutes, depending on experimental conditions).

Starch gel electrophoresis

Starch gel has the advantage of fine texture and absence of protein adsorption, with a molecular sieving effect that retards proteins of larger molecular size. Since native starch does not gel, the starch used for electrophoresis is partially hydrolysed. As the preparation of starch gels is relatively difficult, it has now been superseded by acrylamide gel.

Polycrylamide gel electrophoresis (PAGE)

Acrylamide gel has the advantage over starch in that it is easier to prepare, is more inert, and the pore size can be varied in a controlled manner over a wide range. The technique is similar to starch gel electrophoresis but with a shorter running time.

For the formation of PAGE gel, a solution of acrylamide and bisacrylamide is polymerized, induced by ammonium persulfate (APS) which spontaneously decomposes to form free radicals. Tetramethylethylenediamine (TEMED), a free radical stabilizer, is generally included to promote polymerisation. Acrylamide alone forms linear polymers. The bisacrylamide introduces crosslinks between polyacrylamide chains. The 'pore size' is determined by the ratio of acrylamide to bisacrylamide and by the concentration of acrylamide.

As gel concentration increases, the proportion of bisacrylamide is decreased and the pore size also decreases. Electrophoresis can be carried out in cylindrical glass tubes or by using vertical or horizontal falt-bed equipment. In the simplest form of PAGE, a tube 3 to 10 cm long, in which the gel has been polymerized, is suspended vertically between an upper and a lower buffer reservoir. The buffer, which is the same in both reservoirs and the gel, has a pH (about 9 for protein) such that the macromolecules have net negative charges and hence migrate to the anode in the lower reservoir. Each sample which can contain as little as 10 μg of macromolecular material dissolved in a minimal amount of a relatively dense glycerol or sucrose solution (to prevent it from mixing with the buffer in the upper reservoir) is applied to the top of the gel. A direct current of 300 V is passed through the gel till the components are separated into a series of discrete bands (30–90 minutes). The gel is removed from its holder, and the bands are visualised by an appropriate method. Using this technique a protein mixture of 0.1 to 0.2 mg can be resolved into as many as 20 discrete bands.

Discontinuous (Disc) electrophoresis

In this type of polyacrylamide gel electrophoresis, the given gel slab consist of two separate slabs polymerized one on top of each other, each with a different acrylamide concentration and pH. This type is called discontinuous pH gel electrophoresis, or disc-electrophoresis. The lower concentration gel (stacking gel) is on top of the higher concentration running gel. The system also involves several different buffers. The discontinuity between the gels is maintained on account of the differences in gel structure, pH value of the buffer, ionic strength of the buffer, and the nature of the ions in the gel and electrode buffer.

When current is switched on, the buffer ions from the upper reservoir encounter a pH that is much lower than their pK. They therefore assume their uncharged or, in the case of glycine, zwitterionic form and become electrophoretically immobile. This effect causes the macromolecular ions to approach the running gel as stacks of very narrow (0.01 nm thick) bands or discs that are ordered according to their mobilities. As the macromolecular ions enter the running gel, they slow down as a result of gel filtration effects. This permits the upper reservoir buffer ions to overtake the macromolecular bands and because of the running gel's higher pH, assume their fully charged form as they too enter the gel and electrophoretic separation is achieved. The compactness of the macromolecular bands entering the running gel greatly increases the resolution of the macromolecular separations.

The very large pores needed for the PAGE of large molecular mass compounds (>200 kD) requires gel with very low polyacrylamide concentrations (<2.5%) and hence they are too soft to be usable. This difficulty is overcome by using agarose alone or mixed with polyacrylamide. For example, a 0.8% agarose gel is used for the electrophoretic separation of nucleic acids with molecular masses of up to 50,000 kD.

SDS-PAGE

Sodium dodecyl sulphate (SDS), an anionic detergent, is a strong protein denaturing agent. Most proteins bind SDS in the same ratio, 1.4 g of SDS/g of protein (about one SDS molecule for every two amino acid residues). The large negative charge that the SDS imparts masks the protein's intrinsic charge so that SDS-treated proteins tend to have identical charge-to-mass ratio and similar shapes. Consequently, electrophoresis of proteins in an SDS-containing polyacrylamide gel separates them in order of their molecular masses because of gel filtration effects. The relative mobilities of proteins on such gels vary linearly with the logarithm of their molecular masses. A protein's molecular mass is determined by electrophoresing it together with several "marker" proteins of known molecular mass.

Many proteins contain more than one polypeptide chain, and SDS treatment disrupts the non-covalent interactions between these subunits. Therefore, SDS-PAGE yields the molecular mass of the subunits rather than that of the intact protein unless the subunits are disulfide linked. Mercaptoethanol is usually added to SDS-PAGE gels so as to reduce these disulfide bonds.

Methods of Gel Detection

Gel bands may be detected by staining, radioactive counting or immunoblotting.

Bands resulting from a gel electrophoretic separation can be located by a variety of techniques. Proteins in microgram quantities are often visualised by staining with Coomassie Brilliant Blue. Gel bands with lesser amount of protein can be visualised with silver stain which is 50 times more sensitive. Fluorescamine, an alternative type of protein stain, is a non-fluorescent molecule that reacts with primary amines such as lysine to yield an addition product that is highly fluorescent under UV irradiation. Proteins, DNA, etc., can be detected through the UV absorption of a gel along its length. If the sample is radioactive, the gel may be dried under vacuum after wrapping in cellophane sheet and then damped over a sheet of X-ray film. After some time (few minutes to many weeks depending on the radiation intensity), the film is developed and the resulting autoradiograph shows the position of the radioactive components on development. A gel may also be sectioned widthwise into many slices and the level of radioactivity in each slice determined using a scintillation counter. This latter method yields quantitatively more accurate results than autoradiography.

If an antibody to the protein of interest is available, it is possible to specifically detect this protein on the gel in the presence of many other proteins by immunoblot or western blot.

3. Isoelectric Focussing

A protein has charged groups of both polarities and therefore has an isoelectric point—the pH at which it is immobile in an electric field. If a mixture of proteins is electrophoresed through a solution having a stable pH gradient in which the pH smoothly increases from anode to cathode, each protein will migrate to the position in the pH gradient corresponding to its isoelectric point. Each species of protein is "focussed" into a narrow band about its isoelectric point that may be as thin as 0.01 pH unit.

The pH gradient is stabilised against convection by preparing it in a sucrose concentration gradient by use of a lightly cross-linked polyacrylamide gel in the form of a tube or a thin slab containing urea at a concentration of 6 M. Urea is a powerful protein denaturing agent, but unlike SDS is uncharged and hence cannot directly affect the charge of a protein. Isoelectric focussing can be combined with electrophoresis in an extremely powerful two-dimensional electrophoresis technique.

4. Capillary Electrophoresis

In this technique, electrophoresis is carried out in very thin (1 to 10 μm inner diameter) capillary tubes made of quartz, glass or plastic. Such narrow capillaries rapidly dissipate heat and hence permit the use of high electric fields (100 to 300 V/cm) which reduces separation times to a few minutes. This, in turn, minimises band broadening caused by diffusion, thereby yielding extremely sharp separations. Capillaries can be filled with buffer, SDS-PAGE or polyampholytes. Since capillary electrophoresis can only separate small amounts of material, it is largely limited to use as an analytical technique.

Viva Questions

1. What is electrophoresis? What are the different types of electrophoresis?
2. Name the common applications of electrophoresis.
3. When separated by paper electrophoresis, how many bands are seen in serum proteins? How many bands are seen in lipoproteins?
4. Why does albumin move the fastest in the electrophoretic field?
5. What are the patterns of serum protein separation in nephritic syndrome, multiple myeloma and cirrhosis of the liver?
6. Briefly describe the different types of electrophoresis.

22

BI 11.5; 11.16

Chromatography

Chromatography is a technique of separating closely related compounds from a mixture. In biochemistry, it is used to isolate proteins, peptides, amino acids, carbohydrates, lipids, vitamins and various drugs.

Chromatography involves a **mobile phase** and a **stationary phase**. The mixture of substances to be separated (sample) dissolved in liquid (or gas) is known as the mobile phase. The stationary phase is generally a porous solid matrix or a liquid supported on a solid through which the sample (present in the mobile phase) moves. The interaction between the stationary and mobile phases results in the separation of the mixture in the mobile phase into its various individual components. This involves various physicochemical events such as **partition, ion exchange, molecular sieving, adsorption** and **affinity**. Various chromatographic techniques such as paper chromatography, thin layer chromatography, gas-liquid chromatography and ion exchange chromatography involving the principle of partition are described in this chapter. Paper chromatography is first taken up as a demonstration experiment.

Paper Chromatography (Fig. 22.1)

Paper chromatography is a simple and widely used technique in biochemistry. It was developed by Consden, Gorden and Martin in 1944. It is useful in the separation of closely related compounds from a mixture.

Principle

It is based on the principle of partition of the compounds to be separated between two solvent phases, that is, a stationary phase and a mobile phase.

Methodology

A small amount of the mixture of the compounds to be separated is applied on a strip of special filter paper, a short distance from one end.

The paper serves as a support to hold the stationary phase (water) of the solvent system. The paper is dipped in the solvent system and the solvent is allowed to flow past this part of the application. The solvent system consists of a mixture of N-butanol, glacial acetic acid and distilled water in the ratio 4:1:5. Water is held back by the support (paper) and is called the stationary phase. The organic solvent which moves over the paper is known as the mobile phase. Compounds which are more soluble in the organic solvent move faster.

The relative mobility of the compounds during chromatography is a function of the partition coefficient of the compounds in the two solvent phases.

The partition coefficient is defined as the concentration of solute in the mobile phase divided by the concentration of the solute in the stationary phase. This is denoted by K_d.

$$\text{Partition coefficient } K_d \frac{\text{Concentration of solute in the mobile phase}}{\text{Concentration of solute in the stationary phase}}$$

Thus, similar compounds have different partition coefficients will move to different extents.

The ratio of the distance moved by a compound to the distance moved by the solvent front is known as its Rf value (ratio of fronts).

$$\text{Rf} = \frac{\text{Distance travelled by the compound}}{\text{Distance travelled by the solvent}}$$

Rf values are different for different solvents; for a given solvent, the Rf value is characteristic of the compound and from this observation it is possible to identify unknown substances by their position in relation to the known substances.

Procedure

1. Take a Whatman No.1 chromotography paper of dimension 56 × 46 cm. Draw a thin line with a pencil along the width of the paper, leaving 10 cm from one end.
2. Mark five points on the line 8 cm apart and 8 cm away from the edges.
3. Apply 0.01 mL each of the standard amino acids solutions on the five marked spots. Apply small quantities at a time and allow the spot to dry before making the next application. The spot should not exceed a diameter of 0.5 cm.
4. Similarly, apply 0.1 mL of the unknown solution containing a mixture of amino acids. The chromotograpic chamber is made horizontal. The chamber is saturated using the solvent.
5. The Whatman filter paper is then transferred to the chamber by holding it between filter paper clips and anchoring it with the help of a thick glass rod in the trough.

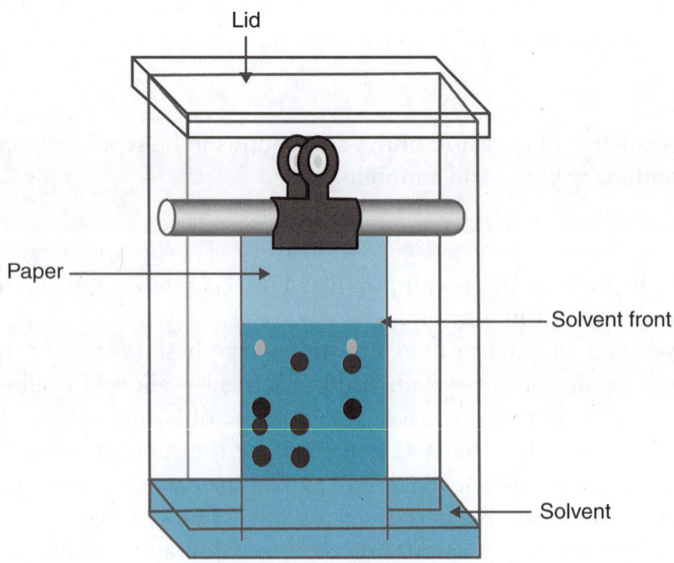

Fig. 22.1 *Paper chromatography*
(https://en.wikipedia.org/wiki/Paper_chromatography)

6. Close the chamber to avoid evaporation of the solvent for about 30 minutes.
7. Fill the trough to nearly three-fourths of its volume with the solvent. Close the chamber and run the chromatography for 16–20 hours.
8. Remove the paper, mark the solvent front and dry the paper using a hair dryer.
9. Spray the paper with ninhydrin staining reagent. Air-dry for about 10–15 minutes and then dry in an oven at 105°C for 5 minutes. The amino acids appear as distinct purple spots (Fig. 22.2).
10. Calculate the Rf values for the amino acids in the given mixture and identify them by comparing with the Rf values of the standards.

Applications

Chromatography is used for

- separation of amino acids, carbohydrates and proteins
- analysis of drugs, vitamins, hormones and brain amines
- qualitative and quantitative analysis of complex mixtures
- estimation of the molecular weight of a protein.

Types of Chromatography

Chromatography can be classified into different types based on its components (mobile and stationary phases) and on the principles it operates upon.

I. Classification of chromatography based on mechanism of separation

Based on the mechanism of separation of molecules in the mixture, chromatography can be classified into the following types:

1. *Adsorption chromatography*: The molecules to be separated in the mobile phase are adsorbed in the resin (stationary phase) packed in a column. Affinity and HPLC chromatographies belong to this category.

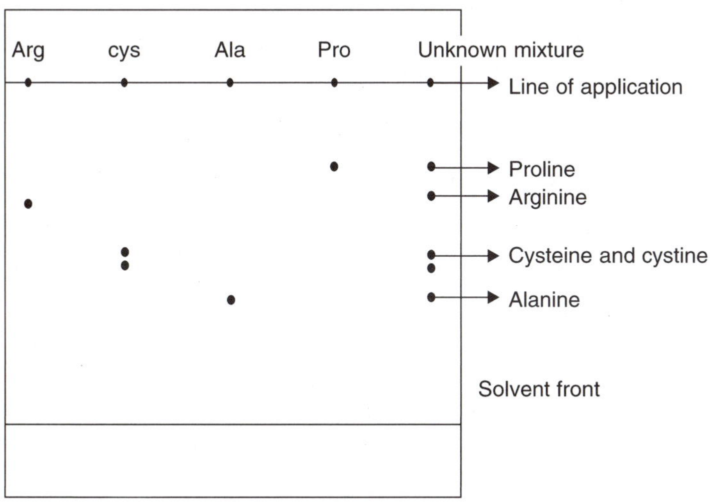

Fig. 22.2 *Chromatographic separation of amino acids*

2. *Partition chromatography*: In this type of chromatography, particles of the solute get partitioned based on their solubility properties. For example, in paper chromatography, the mobile phase is a solvent and water absorbed on paper serves as the stationary phase. The particles are separated out by the process of partition based on their solubility between the liquid mobile phase and the liquid (water) stationary phase.
3. *Ion-exchange chromatography*: In this process, the separation of ions and polar molecules takes place on the basis of their affinity to ion exchangers. The ions from the sample to be separated gets attached to the resin, replacing the ones originally attached to it.
4. *Size-exclusion chromatography*: The separation of molecules occurs based on their size as they pass through a resin packed in a column. It is also known as gel-permeation chromatography.

II. Classification of chromatography based on mobile phase

Based on the type of mobile phase, chromatography is categorised into three categories. The mobile phase can be liquid, gas (never solid) and supercritical fluid.

1. *Liquid chromatography*: In this type, the mobile phase is a liquid. It includes normal and reverse phase chromatographies, hydrophilic and hydrophobic interaction chromatographies, ion-exchange chromatography and affinity chromatography.
2. *Gas chromatography*: In this variety, gas serves as the mobile phase. Gas chromatography falls in this category.
3. *Supercritical fluid chromatography*: A supercritical fluid is a substance above its critical temperature and critical pressure, where distinct liquid and gas phases do not exist. Supercritical carbon dioxide (CO_2) is used as the mobile phase in this type of chromatography.

III. Classification of chromatography based on stationary phase

The stationary phase can either be a solid or a liquid but never a gas. The chromatography is classified on the basis of the separation belt (the resin on which separation occurs).

1. *Planar chromatography:* The resin on which separation occurs is a planar surface, such as in paper chromatography and thin layer chromatography.
2. *Column chromatography:* The stationary phase resin occurs in a column, such as in ion-exchange, HPLC, gas chromatography, etc. The column can be either packed bed or open tubular.

Thin Layer Chromatography

This is an improvement over paper chromatography. In this technique, separation of molecules from a mixture occurs based on the polarity of molecules.

Procedure

To perform TLC, we need a TLC plate which can be made in the laboratory or bought readymade from the market. A glass slide (its size can be varied according to the number of compounds to be separated) is taken and on it a thin layer of silica gel is applied. (In readymade TLC plates, an aluminium base is used in place of the glass slide. As the layer of silica gel is distributed more uniformly in thickness in the readymade plates, they perform better than lab-made glass slides.) Use any end of the TLC plate as the base side and draw a thin line with a pencil 1 cm above the edge of the plate for applying the sample(s). Ink should not be used as it will spread on the plate during run. After applying the sample, place the TLC plate in a chamber/beaker containing the solvent (mobile phase) such that the sample application line lies above the solvent level (for, if it is immersed in the solvent, the sample will dissolve in it). Before placing the TLC in the beaker/ chamber, however, it has to be ensured that the chamber is saturated with the solvent. For this, the beaker is kept closed using a lid for about 20 minutes or so after the solvent has been poured in, so that the solvent vapour gets distributed all over the beaker.

After keeping the TLC plate in the beaker, close it with a lid. The solvent allows the sample(s) to move upwards (through capillary action) with varying speeds based on the degree of their polarity. Since silica gel is polar in nature, the more polar molecules move slower on the plate and the less polar molecules move faster. Simultaneously, the solvent moves on the TLC plate at its own speed (Fig. 22.3). If the compound separated is a coloured one, it is displayed clearly on the plate, but if it is colourless, different methods are employed to visualise it. Polyphenol compounds are visualised by UV lamp through fluorescence. Other colourless compounds are spotted by different means such as acid treatment, iodine vapour (compounds containing double bonds), vallinin (alcohol, aldehyde and ketone compounds), ninhydrin (for amino acids), etc. If no compound is visible, that means inadequate sample is applied. If the rim shows streaks of material, it indicates overload of sample. If the solvent (mobile phase) run is not equal, it indicates that the silica gel thickness is not uniform.

The distance travelled by the separated compound and the solvent are measured, and Rf is estimated as

$$\text{Rf} = \frac{\text{Distance travelled by the separated compound}}{\text{Distance travelled by the solvent (mobile phase)}}.$$

Each compound has a specific Rf value. By comparing the Rf of the newly separated compound (from the mixture) with its reference value, the compound can be identified.

Gas–Liquid Chromatography

This is used for separating and identifying volatile compounds of fatty acids, hydrocarbons, sterols, amino acids and lipids in a sample, even when the sample is available only in trace quantities. Since volatility is an essential criterion for this method, the compounds should be non-polar or made non-polar through chemical derivatization. The vaporized mixture of volatile

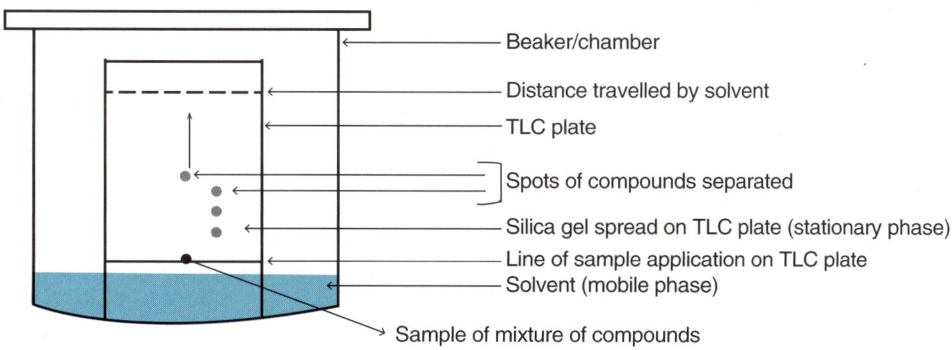

Fig. 22.3 *Diagrammatic representation of TLC*

compounds are carried by an inert gas such as He, Ar or N_2 (mobile phase) through a long hot tube containing a column of inert fibre brick or diatomaceous earth particles impregnated with a non-volatile liquid like silicone oil or grease (the stationary phase). The volatile components separate from each other according to their partition coefficients between the mobile gaseous phase and the stationary liquid phase. As the compounds leave the column, they pass through a detector which is linked via an amplifier to a chart recorder which records a peak corresponding to the relative amounts of the different chemicals in the sample.

Ion Exchange Chromatography

This method separates amino acids, oligonucleotides, peptides or proteins, depending on their acid–base properties and charges. It is a form of chromatography in which the net positive or negative charge that macromolecules or biomolecules carry is exploited for its purification. The charged molecules in the sample are separated by the electrostatic forces of attraction between them and the ionic resin (ion exchanger) through which they pass. Ion exchangers are insoluble polymers like cellulose, agarose or polystyrene beads on whose surface either positively charged (anion exchanger) or negatively charged (caption exchanger) groups chemically attach. Cation exchangers possess negatively charged groups and attract positively charged molecules. Anion exchangers have positively charged groups and attract negatively charged molecules.

$$\text{Resin-H}^+ + \text{Na}^+ \rightarrow \text{Resin-Na}^+ + \text{H}^+ \text{ (cation exchanger)}$$

$$\text{Resin-OH}^- + \text{Cl}^- \rightarrow \text{Resin-Cl}^- + \text{OH}^- \text{ (anion exchanger)}$$

The entire procedure relies on the principle that the charged biomolecules adsorb on ion exchangers reversibly and the bound molecule is released or eluted by changing the ionic environment. The functional groups attached to anion exchangers are QAE and DEAE (quaternary amino ethyl or diethyl amino ethyl), while in cation exchangers these are SP, P or CM (sulpho propyl or phosphor or carboxy methyl). In general, Dowax type materials (polystyrene beads) are used for the separation of low molecular weight molecules and agarose or cellulose type materials are used for the separation of lower molecular weight compounds.

Choice of ion exchanger

Many substances, e.g., proteins, carry both positive and negative charges and the net charge depends on their pH. As a rough guide, if a protein is stable at pH values above the isoelectric point, an anion exchanger should be used; if it is stable below the isoelectric point, a cation exchanger should be used.

Choice of pH, buffer and ionic conditions

The pH of the buffer used should be at least one unit above or below the isoelectric point of the compounds to be separated. In general, cationic buffers such as Tris, pyridine and alkylamines are used with anion exchangers, and anionic buffers such as acetate, barbiturates and phosphates are used with cation anion exchangers. To get good separation or resolution of peaks, the method of choice is to use gradual increase of pH and or ionic strength. This is achieved by using a gradient mixer which is very easy to operate in a laboratory. While choosing the buffers, it is better to use volatile buffer because the recovery of the pure material can be achieved through lypophilization.

Gel Exclusion (Permeation) Chromatography

This technique makes use of the fact that molecules get sieved on the basis of their molecular size and shape when passed through a porous material. It has wide spread applicability in molecular biology for the separation of molecules based on size.

The porous materials which are used are a group of polymeric organic compounds that possess a three-dimensional network of pores. In practice, a column of such a porous material is made and a solution containing molecules of various dimensions is made to pass through the column. Larger molecules are completely excluded from the pores and pass through the interstitial space, while the smaller molecules are distributed between the solvent inside and outside the molecular sieve and then pass through the column at a slower rate. Hence, the molecules are eluted from the column in the order of decreasing molecular weight, i.e., large (excluded) molecules are eluted first. This method is extensively used in separating immunoglobulins from other proteins, nucleic acids, viruses and ribosomes.

A buffered solution of the mixture is made to flow under gravity through the column of porous gels. The gels used as molecular sieves consist of cross-linked polymers of dextran (Sephadex), agarose (sepharose, biogel), polyacrylamide (biogel P) and polystyrenes (bio-beads S). These gels swell in water and few organic solvents but not in non-polar solvents. By using suitably modified gels (LH-20 or 40), both polar and non-polar solvents can be used as mobile phase for the purification of lipids, fatty acids and steroids.

Large solute molecules flow at faster rates down the column because they cannot penetrate the beads, but small molecules penetrate the beds through the pores and consequently get considerably retarded. The degree of penetration and retardation is inversely proportional to the particle size. After elution, the molecules or particles can be collected in different fractions of the elute in the order of their decreasing size.

Affinity Chromatography

Affinity chromatography is a type of chromatography that makes use of a specific affinity between a substance to be isolated and a molecule to which it can specifically bind as a ligand. It is used for the purification of enzymes, nucleotides, nucleic acids, immunoglobulins, membrane receptors and even whole cell and cell fragments. The method exploits the unique property of extremely specific biological interactions to achieve separation and purification. For isolating an antigenic protein, the ligand used is a specific antibody molecule that react with the antigen; for an enzyme, the ligand is its substrate or competitive inhibitor, etc.

The column material is synthesised by covalently coupling the ligand to an insoluble matrix. When a solution containing the desired protein is passed through a column of the ligand-bound gel, the ligand molecule binds only with the specific protein molecule; the other proteins pass through the column without being adsorbed. After washing out all unbound substances from the column, a buffer solution of specific pH and ionic strength is percolated through the column to release the ligand-bound protein in the elute by dissociating the non-covalent bonds between the adsorbed protein and the bound ligand.

High Performance (Pressure) Liquid Chromatography (HPLC)

The resolving power of a column increases with the length of the column and the number of equilibrations per unit length of the column. The number of equilibrations is related to the surface area of the stationary phase. The flow rate of a column drops as the particle size decreases even though the number of equilibrations is more. This problem can be overcome if a pressure difference is established between the top and bottom of the column (via pumps) in order to force the liquid into the bed. In HPLC, a pump forces the solvent through a column at a high pressure (up to 10,000 Psi), making it a powerful and versatile form of chromatography. The use of high pressure with very fine particles to maintain an adequate flow rate is characteristic of HPLC. The principles of adsorption, partition, ion exchange, exclusion and affinity used in the technique results in faster and better resolution of the components of the mixture.

Viva Questions

1. What is chromatography? Name the different types of chromatography.
2. What are the stationary and mobile phases in paper chromatography?
3. What is Rf value?
4. What is the advantage of thin layer chromatography over paper chromatography?
5. What are the common applications of chromatography?
6. Briefly describe the different types of chromatography.

23

BI 11.16

Radioimmunoassay (RIA) and Enzyme-linked Immunosorbant Assay (ELISA)

Immunochemical techniques are usually employed to detect or quantitate the antigen or antibody. RIA and ELISA are two important immunoassay techniques used to measure hormones, drugs, tumour markers and antigens which occur in micro quantities in biological samples.

Radioimmunoassay (RIA)

This technique was first developed by Benson and Yallow in 1959 for estimation of insulin. Now it is used extensively in biochemistry and clinical medicine for estimating a variety of compounds that occur in as low as nanogram to picogram quantities.

Like any immunoassay technique, it involves a reaction between an antigen and its specific antibody. RIA utilises radioactive isotopes.

Principle

RIA is primarily based on the competition between unlabelled antigen (Ag) in the samples or standards and radio-labelled antigen (Ag*) for the limited number of available binding sites on the specific antibody (Ab). At the end of incubation, the bound antibody and free antigen are separated by the addition of the precipitating agent. Antigen concentrations of the samples are quantitated by measuring the radio activity associated with the bound fraction of samples and standards in a radioactive counter. As the concentration of unlabelled antigen (Ag) increases, the amount of labelled antigen–antibody (Ag*–Ab) decreases. Therefore, the radioactivity count is inversely proportional to the unlabelled antigen concentration.

In India, the kit for RIA can be obtained from the Board of Radiation and Isotope Technology, Mumbai, Dr Reddy's Laboratory, Hyderabad and Spectral Diagnostics, Delhi. Manufacturer's instructions in the kit can be followed to assay the analyte in the given biological sample.

The method is highly sensitive, and hence extremely low concentrations (up to picograms) of the antigen can be measured. It is highly specific in its action. On the other hand, the reagents and equipment are expensive. The shelf-life of reagents is short and hence cannot be stored for long. The assay is of long duration. Adequate safety measures must be taken while handling and disposing radioactive materials to avoid radiological hazards.

Applications

A large number of compounds which occur in minute concentrations in biological fluids can be accurately quantitated.

1. Hormones such as thyroid (T_3, T_4 and TSH), growth hormone, insulin, estradiol, FSH, LH, prolactin, progesterone, testosterone, aldosterone, ACTH, cortisol, PTH and renin can be accurately measured.
2. Tumour markers such as prostate specific antigen (PSA), alpha-fetoprotein (AFP), and so on can be detected.
3. All vitamin levels can be measured.
4. Therapeutic monitoring of drugs (for example, digoxin) can be done.

Enzyme-linked Immunosorbant Assay (ELISA)

ELISA was introduced in 1971 and since then, it has been preferred over RIA. This is because it is considered more sensitive than RIA, simple and easy without sophisticated instrumentation (hence can be used even in small laboratory set-ups) and, most importantly, there is no exposure to radioactivity.

In this technique, enzymes are used to label the antigens or antibodies. ELISA can be performed using two methods:

1. Single antibody method
2. Double antibody method

Single antibody method

This is commonly known as *competitive method*. In this technique, a known amount of enzyme-labelled antigen and an unknown amount of unlabelled antigen (in the patient sample) mixture are allowed to react with a specific antibody fixed on an inert solid.

There is competition between the labelled antigen and unlabelled antigen for binding with the limited number of antibodies available. Wells are washed after the incubation period. During washing, the unbound antigens are washed off with buffer. After washing the unbound antibody–enzyme conjugate with buffer, the enzyme substrate is added and enzyme activity is measured (by measuring the product formed colorimetrically). The enzyme activity measured is directly proportional to amount of the labelled antigen and inversely proportional to the amount of the unlabelled antigen in the test sample.

Double antibody method

This is popularly known as *sandwich method*. In this technique, the unknown antigen in the test sample is allowed to bind with the specific antibody attached covalently to a solid support like a thin sheet polyvinyl chloride. Now a second antibody labelled with the enzyme is added. This antibody binds with the already bound antigen, forming an antibody–antigen–antibody complex. The antigen is now in a state of being 'sandwiched' between two antibodies. After washing off the excess antibodies, the enzyme substrate is added. Enzyme activity is measured by measuring the product formed colorimetrically. The enzyme activity is directly proportional to the amount of antigen present in the test sample. For each molecule of antigen that binds in the final complex, there will be thousands of product molecules produced which cause amplification. This amplification effect makes ELISA a highly sensitive immunoassay.

Immunochemical techniques are usually employed to detect or quantitate the antigen or antibody. RIA and ELISA are two important immunoassay techniques used for measuring hormones, drugs, tumour markers and antigens that occur in micro quantities in biological samples. Like any immunoassay technique, both involve a reaction between an antigen and its specific antibody. RIA utilises radioactive isotopes.

ELISA was first introduced in 1971 and since then, it has been preferred over RIA. This is because it is considered more sensitive than RIA, simple and easy to use without the need for sophisticated instrumentation (hence useable even in small laboratory setups) and, most importantly, without the risk of exposure to radioactivity.

There are four major types of ELISAs: direct, indirect, competitive and sandwich (Fig. 23.1).

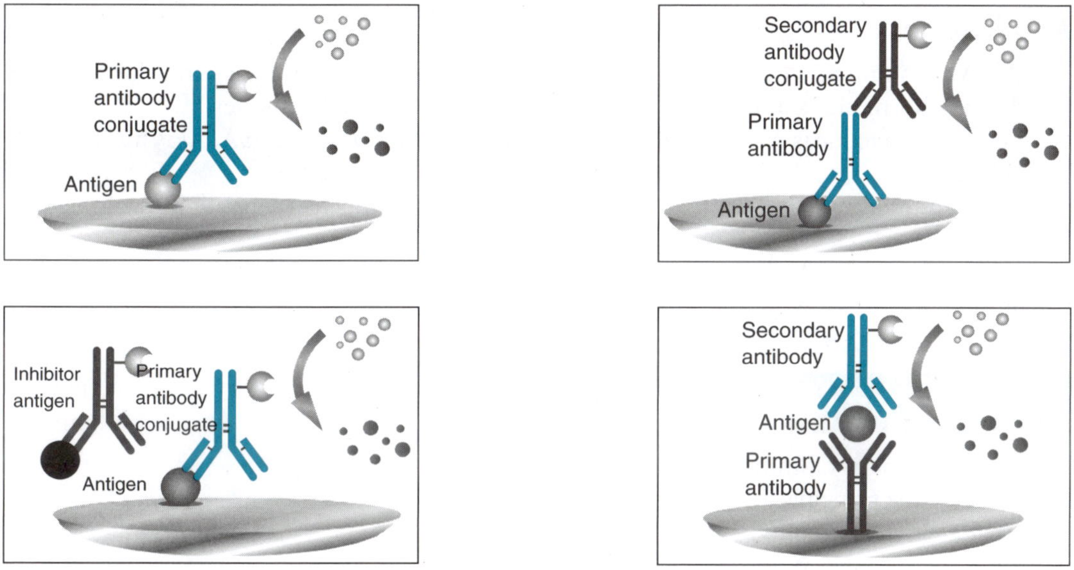

Fig. 23.1 *Types of ELISA: (a) direct, (b) indirect, (c) competitive and (d) sandwich*

Direct ELISA

In a direct ELISA, the antigen is bound to the bottom of the microplate well; it is then bound by an antibody that is specific to the antigen and also conjugated to an enzyme or other molecule that enables its detection.

Indirect ELISA

In an indirect ELISA, the antigen is bound to the bottom of the microplate well. Then an antibody specific to the antigen is added. A secondary antibody, conjugated to an enzyme or other detection molecule, is then bound to the first antibody.

Competitive ELISA

In a competitive ELISA, a reference antigen is bound to the bottom of microplate wells. Sample plus antibody are added to the wells, and if there is antigen present in the sample, it competes with the reference antigen for binding to the antibody. Unbound material is washed away.

The more antigen there is in the sample, the less the antibody that ends up bound to the bottom of the wells by the reference antigen and the lower the signal.

Sandwich ELISA

For the sandwich ELISA, two antibodies specific to two different epitopes on the target antigen are used. The capture antibody is bound to the bottom of the microplate well and binds one epitope of the antigen. The detection antibody binds to the antigen at a different epitope and is conjugated to an enzyme that enables detection. (If the detection antibody is unconjugated, then a secondary enzyme-conjugated detection antibody is required).

Applications

1. ELISA is commonly used to measure hormones in serum such as T_3, T_4 and TSH, insulin, pituitary hormones (for example, FSH, LH, TSH, etc.) and sex hormones.
2. It is also used to measure the levels of tumour markers in serum like PSA, AFP, hCG, CA125 and CEA.
3. It is used to detect bacterial toxins, viruses and surface antigens (hepatitis B) in infectious diseases.
4. ELISA is used for the assay of auto antibodies such as anti-DNA and ANA (antinuclear antibody).

Viva Questions

1. What is the basis of immunochemical techniques?
2. What are the principle and applications of RIA?
3. What are the different types of ELISA?
4. What are the applications of ELISA?
5. What is indirect ELISA method?
6. Why is ELISA commonly preferred over RIA?

24 Estimation of pH and Preparation of Buffers

BI 11.2

All the biochemical reactions in the body take place in an aqueous environment and most of these reactions are catalysed by enzymes. Enzymes are optimally active at a particular H^+ ion concentration. Water (H_2O) is dissociated to H^+ and OH^- ions. In health, the H^+ ion concentration varies very little in any biochemical fluid or environment. The variation is arrested by some bases or acids which respectively absorb or donate H^+ ions depending on the situation. This phenomenon is called as buffering.

H^+ ions can be measured and expressed as mol/L. But there is a more convenient way to express H^+ ion concentration, i.e., pH, which is defined as the negative logarithm of the number of hydrogen ions (in moles) per litre of liquid. (The letter 'p' in pH stands for the negative common (base ten) logarithm, while 'H' stands for the element hydrogen.).

$$pH = -\log(H^+)$$

In aqueous solution, the pH ranges from 0 to 14. Molar concentration of H^+ or OH^- ion in pure water is 1×10^{-7}/L. The ion product of water is 1×10^{-14} molar^2/L^2.

In water that is neutral (at neutral point),

$$[H^+] = [OH^-] = 10^{-7} \text{ mol/L}.$$

So, pH at neutral point is 7.

$$-\log (10^{-7}) \text{ mol/L} = 7.$$

A pH more than 7 indicates that the solution is alkaline and a pH less than 7 that the solution is acidic.

A pH of 0 would be given by 1 M HCl.

A pH of 14 would be given by a 1 M NaoH.

The pH of blood is 7.4 (slightly alkaline); the pH of gastric juice it is 1.5 to 2.0 (highly acidic).

Many biochemical substances possess functional groups that are weak acids or weak bases, e.g., –COOH, NH_2, and phosphates present in proteins, nucleic acids, coenzymes, and intermediary metabolites. Their dissociation behaviour influences pH and the structure and function of the entire molecule/molecules.

Consider an acid HA, whose dissociation can be given by

$$HA \rightleftharpoons H^+ + A^-.$$

The relative strength of an acid (HA) is expressed by its dissociation constant K_a.

$$K_a = \frac{[A^-]\,[H^+]}{[HA]}.$$

The quantities in square brackets represent the concentrations of the product of dissolution at equilibrium.

The numerical value of K is a negative exponential number, and so it is usually expressed as pK, where,

$$pK = -\log K.$$

The lower the pK, the stronger is the acid.

Relationship of pH to pK

Such a relationship is given by the Henderson–Hasselbalch equation:

$$pH = pK_a + \log\frac{[\text{salt}]}{[\text{acid}]} = pK_a + \log\frac{[A^-]}{[HA]}.$$

Derivation

$$K_a = \frac{[H^+]\,[A^-]}{[HA]} \Rightarrow [H^+] = K_a\frac{[HA]}{[A^-]}$$

or $$-\log\,[H^+] = -\log\left[K_a\frac{[HA]}{[A^-]}\right]$$

or $$pH = pK - \log\frac{[HA]}{[A^-]}$$

$$= pK + \log\frac{[A^-]}{[HA]}.$$

Application: To calculate pH, pK or molar ratio of H^+ donor and acceptor.

Example: Calculate the pH of a mixture of 0.1 M acetic acid and 0.2 M sodium acetate (pK of acetic acid is 4.76).

Ans: $$pH = 4.76 + \log\left(\frac{0.1}{0.2}\right)$$

$$= 4.76 + 0.301 = 5.06.$$

Determination of pH Using a pH Meter

There are two methods of measuring pH: (i) the colorimetric method which uses indicator solutions or indicator papers to produce colour depending on the pH of the sample and (ii) the electrochemical method using a pH meter, which is more accurate than the colorimetric method.

Principle of pH meter

The pH of a solution can be determined easily by measuring the hydrogen ion (proton) concentration in it using a device known as the pH meter. The device is an electrochemical cell containing a pH (proton) sensitive electrode made of glass, a reference electrode (usually silver chloride or calomel), a temperature sensing element and the liquid (test solution) whose pH is being measured (Fig. 24.1). The design of the pH electrode is such that it allows hydrogen ions in the test solution to migrate through a selective barrier to produce a potential (voltage) difference proportional to the pH of the solution. The reference electrode is designed to maintain a constant potential at any given temperature. It provides a known reference potential for the pH electrode. The difference in the potentials of the pH and the reference electrodes generates a signal in the millivolt range, proportional to the pH of the sample solution. The proportionality constant depends on temperature, so a temperature sensor is necessary in the setup. The commercially available pH electrodes are mostly combination electrodes, wherein both the glass H^+-ion-sensitive measuring electrode and the reference electrode are placed in one housing. It is calibrated against buffer solutions of known pH.

Procedure

Measuring pH is about measuring the acidity (pH 0–7) or alkalinity (pH 7–14) of an aqueous sample. The meter which performs this function works like a volt meter (Fig. 24.1), measuring the potential difference between the known reference electrode and the measuring pH electrode. The measuring electrode responds to the concentration of the hydrogen ions in the sample solution (whose pH is being measured) to produce a chemical signal. This signal is then converted into pH values and displayed on the electronic control unit.

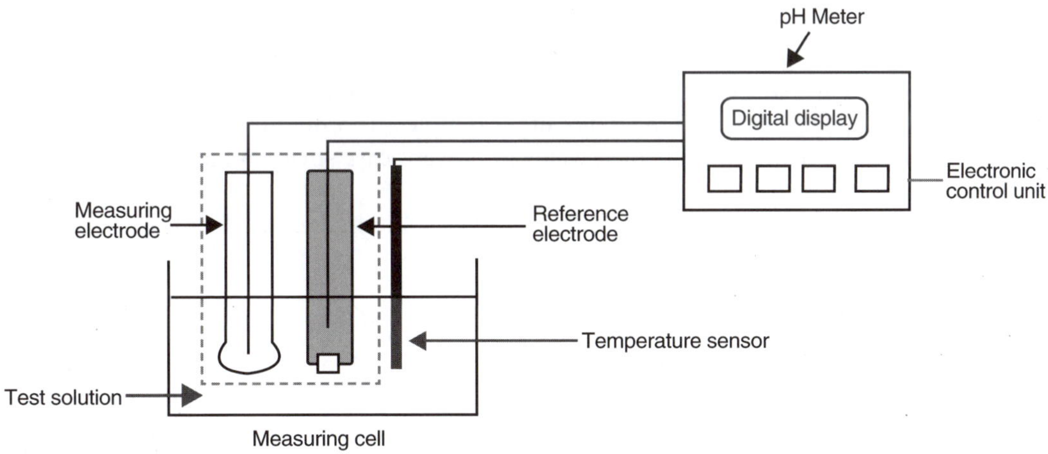

Fig. 24.1 *Schematic diagram of a pH meter (the dashed unit in the measuring cell is generally packaged as a combination electrode).*

Before using the pH meter, it must be calibrated with standard buffer solutions of known pH values. In the process described below, three buffer solutions with pH values 7, 4 and 10 are used for calibration. To make the pH measurement, the electrode is immersed in the sample solution till a steady reading is obtained on the display.

pH Calibration

1. To start operating the pH meter, turn on the 'Power' button and set the mode to 'pH' which will appear on the LCD display.

2. Plug in the pH probe and the temperature probe into the unit.
3. Choose a fresh pH buffer for calibration, say with pH 7.0. (The pH buffer must be at the same temperature as the testing solution.) Rinse the pH electrode first with distilled water and then with the standard pH 7.0 buffer used for the calibration.
4. Now, dip the pH electrode into the pH buffer (pH 7.00), stirring the buffer with a magnetic bar.
5. Start the calibration/standardisation button on the control unit and set the buffer pH to 7.00. When the reading is stable, press the 'Enter' button.
6. Rinse the pH electrode with distilled water and then with the standard buffer to be used for next calibration (i.e., pH 4.00). Repeat the procedures as above, i.e., dip the pH electrode into the standard pH 4.0 buffer solution with stirring and lock the display at pH 4 by pressing the 'Enter' button.
7. Repeat the above steps for further calibration with pH 10.00 buffer.
8. After completing the 3-point calibration, the pH meter will automatically return to measurement mode.

It is recommended that calibration be performed the at the beginning of each day for accurate results of the pH meter.

pH Measurement

1. Ensure that the pH meter is set on the pH measurement mode. Dip the pH electrode into a testing solution and stir the solution with a magnetic rod.
2. When the pH measurement is done, record the value by writing it down or pressing the memory button on the meter (if available) to store the reading in memory.
3. Repeat the steps for measuring the PH of other samples. Thoroughly rinse the pH electrode between measurements with distilled water to prevent carry over contamination from the tested solutions.
4. Thoroughly rinse and store the electrode in storage solution once all the measurements are completed.

Applications of pH measurement

Estimation of pH value is useful for diagnosing disease conditions. The pH of body fluids in living organisms is tightly regulated through acid–base balance. The pH of blood is slightly basic with a value of 7.365. The pH differs in different cellular compartments, body fluids and organs. Enzymes and other proteins function at an optimum pH range of 6–8; they get denatured and lose their function outside this range.

Besides applications in medicine, the measurement of pH values is of significance in various agricultural, industrial, pharmaceutical and environmental applications. Pure water is neutral with a pH of 7.00. When chemicals are mixed with water, the mixture can become either acidic or basic, which can be ascertained by measuring its pH.

Buffers

A buffered solution is one which resists a change in pH when small amounts of acid or base are added to it. Such a solution is prepared from a mixture of weak acid or base with its salt (conjugate base or acid).

Examples of buffer systems: Carbonic acid/bicarbonate; dihydrogen phosphate/monohydrogen phosphate; proteins/proteinate.

The principle of the buffering action is the common ion effect. The addition of OH^- (base) results in its combination with H^+ buffer, leading to the lowering of H^+ concentration and further ionization of the weak acid. Thus OH^- is neutralised and H^+ concentration is maintained. With addition to H^+ (acid), H^+ dissociation is suppressed and H^+ concentration is maintained by backward dissociation reaction. Buffers of weak base and its salt maintain pH by a similar mechanism.

The buffering capacity is maximal when the salt and acid concentrations are equal (i.e., pH = pK). The buffers mentioned in the examples above are present in different body fluids, helping the regulation of pH. Buffers are used in clinical chemistry for enzymatic estimations.

Buffer solutions

When a strong acid or alkali is added to a litre of water (pH 7.00), its pH changes completely. On the contrary, if the same strong acid or alkali are added to a buffer solution, the change in its pH is only marginal. Observe the following example:

Solution	Initial pH	Add	Final pH	Change in pH
Water	7.0	0.10 mol HCl	1.00	−6.00
Water	7.0	0.10 mol NaOH	13.0	+6.00
Buffer $[CH_3COOH] = 1.0$ M $[CH_3COOH^-] = 1.0$ M	4.74	0.10 mol HCl	4.66	−0.08
Buffer $[CH_3COOH] = 1.0$ M $[CH_3COOH^-] = 1.0$ M	4.74	0.10 mol NaOH	4.83	+0.09

From the readings above, it is becomes apparent that when a small quantity of acid is added to pure water, its pH drops dramatically; likewise, when a small quantity of base is added to pure water, its pH rises dramatically. On the contrary, when the small quantity of acid/base is added to a buffer solution, the pH drops/rises only very slightly. To summarise, a buffer solution possesses the ability to resist an appreciable change in H^+ ion concentration when small quantities of acid or base are added to it.

Preparation of Buffers

A buffer solution must be able to partially neutralize both acids and bases that are added to it. In order to do this, it must contain relatively high amounts of both base and acid, which is possible only when both the base and acid in it are weak. A buffer solution consists of a *weak conjugate acid–base pair* in which both the acid and the base have relatively high concentrations. A common example is a solution with 1 M CH_3COOH (weak acid) and 1 M CH_3COO^- (weak base).

An equilibrium is established when the concentration of CH_3COOH and CH_3COO^- are both 1 M.

$$\underset{(1\,M)}{CH_3COOH} + H_2O \rightleftharpoons H_3O^+ + \underset{(1\,M)}{CH_3COO^-}$$

In the above equation the concentration of hydronium ion (H_3O^+) is low. 1 M of acetic acid is able to neutralise a small quantity of strong base added to this solution, and 1 M of acetate ion is able to neutralise a small quantity of strong acid added to this solution. To prepare buffer solutions, molecular acids can be added directly. Concentrated acetic acid (ethanoic acid) can be diluted to 1 M and soluble ionic salts used as a source of ions; 1 mol of solid sodium acetate (sodium ethanoate, i.e., $NaCH_3COO$) can be added, making one litre of solution. Sodium acetate is dissociated to sodium and acetate ions.

$$NaCH_3COO \rightarrow Na^+ + CH_3COO^-$$

In the above, Na^+ is not a part of buffer; only CH_3COO^- is a part of it. The buffer prepared with 1 M CH_3COOH (weak acid) and 1 M $NaCH_3COO$ (salt of its conjugate base) is an example of *acidic buffer*.

Another example of a buffer solution is the weak base ammonia (NH_3) and its conjugate acid, the ammonium ion (NH_4^+).

$$\underset{(1\,M)}{NH_3} + H_2O \rightleftharpoons \underset{(1\,M)}{NH_4^+} + OH^-$$

Molecular bases such as NH_3 can be added directly. Soluble ionic salts are sources of ions; 1 mol of solid ammonium chloride (NH_4Cl) can be added.

$$NH_4Cl \rightarrow NH_4^+ + Cl^-$$

In the above equation, Cl^- is not part of the buffer; only NH_4^+ ion is part of it. The buffer prepared with 1 M NH_3 (weak base) and 1 M NH_4Cl (salt of its conjugate acid) is an example of *basic buffer*.

In some buffer solutions, both the weak acid and weak base are ions. For example, a solution containing 1 M of $H_2PO_4^{2-}$ (dihydrogen phosphate) and 1 M of HPO_4^{2-} (monohydrogen phosphate):

$$\underset{\substack{(1\,M)\\(\text{weak acid})}}{H_2PO_4} + H_2O \rightleftharpoons H_3O^+ + \underset{\substack{(1\,M)\\(\text{weak acid})}}{HPO_4^{2-}}$$

Any soluble ionic compounds such as KH_2PO_4 serves as a source of $H_2PO_4^-$ and K_2HPO_4 serves as a source of HPO_4^-.

$$\underset{(1\,M)}{KH_2PO_4} \rightarrow K^+ + H_2PO_4^-$$

$$\underset{(1\,M)}{K_2HPO_4} \rightarrow 2K^+ + HPO_4^{2-}$$

It may be noted that strong acids and their conjugate bases or strong bases and their conjugate acids are not used to prepare buffer solutions. *In order to be an effective buffer, a solution must contain both a weak acid and a weak base. The concentration of weak acid and the concentration of weak base should be relatively high and should be equal or close to each other.*

The composition of various buffers is given below.

Phosphate Buffer (0.1N)

PH value	0.1N KH_2PO_4 (in mL)	0.1N Na_2HPO_4 (in mL)
5.8	92.0	8.0
6.2	81.4	18.6
6.5	68.2	31.8
7.0	38.9	61.1
7.4	19.2	80.8
7.9	6.8	93.2

(KH_2PO_4: Potassium dihydrogen phosphate; Na_2HPO_4: Disodium hydrogen phosphate)

Barbitone Buffer, pH 7.4

Sodium diethyl barbiturate ($C_8H_{11}O_3N_2Na$) (57 mmol/L)	11.74 g
Hydrochloric acid (100 mmol/L)	430 mL

Barbitone Buffered Saline, pH 7.4

Sodium chloride (NaCl)	5.67 g
Barbitone buffer, pH 7.4	1 mL

Before use, dilute with an equal volume of 0.9 g/L NaCl.

Barbitone Buffered Saline, pH 9.5

Sodium diethyl barbiturate ($C_8H_{11}O_3N_2Na$) (98 mmol/L)	20.2 g
Hydrochloric acid (100 mmol/L)	20 mL
Sodium chloride (NaCl)	5.67 g

Before use, dilute with an equal volume of 0.9 g/L NaCl.

Barbitone–Bovine Serum Albumin Buffer, pH9.8

Hydrochloric acid (100 mmol/L)	5 mL
Sodium diethyl barbiturate ($C_8H_{11}O_3N_2Na$) (54 mmol/L)	10.3 g
Sodium chloride (NaCl) (102 mmol/L)	6 g
Sodium azide (31 mmol/L)	2 g
Bovine serum albumin (BSA)	5 g
Water to make volume up to 1 L	–

Dissolve the reagents in 900 mL of water. Adjust the pH to 9.8 with 5 mol/l HCl. Make up the volume to 1 L with water. Store at 4°C.

Citrate–Saline Buffer, pH 6.0

Trisodium citrate ($Na_3C_6H_5O_7.2H_2O$) (5 mmol/L)	1.5 g
Sodium chloride (NaCl) (96 mmol/L)	5.6 g
Barbitone buffer, pH 7.4	200 mL
Water	800 mL

Glycine Buffer, pH 3.0

Glycine (NH_2CH_2COOH) (82 mmol/L)	6.15 g
Sodium chloride (NaCl) (82mmol/L)	4.8 g
Water	820 mL
0.1mol/L HCl	180 mL

(HCl: Hydrochloric acid; NaCl: Sodium chloride)

3-[4-(2-Hydroxyethyl)-1-Piperazinyl]-Propanesulfonic Acid (HEPPS) Buffer, pH 6.6

3-[4-(2-Hydroxyethyl)-1-piperazinyl]-propanesulfonic acid (100 mmol/L)	23.83 g
Sodium hydroxide (NaOH) (1 mol/L)	Approximately 1 mL
Dimethyl sulfoxide	25 mL
Water to make volume up to 1 L	–

Dissolve in 100 mL of water. Add a sufficient volume of 1 mol/L NaOH (1 mL) to adjust the pH to 6.6.

If the buffer is intended for use with Romanowsky staining, then add 25 mL of dimethyl sulfoxide (DMSO). Make up the volume to 1 L with water.

3-[4-(2-Hydroxyethyl)-1-Piperazinyl]-Propanesulfonic Acid (HEPPS) Buffer, pH 7.6

3-[4-(2-Hydroxyethyl)-1-piperazinyl]-propanesulfonic acid (20 mmol/L)	4.76 g
Sodium hydroxide (NaOH) (1 mol/L)	8 9 g
Sodium chloride (NaCl)	–
Water to make volume up to 1 L	–

Dissolve in approximately 100 mL of water. Add a sufficient volume of 1 mol/L NaOH to adjust the pH to 7.6. Make up to the volume to 1 L with water.

Imidazole Buffered Saline, pH 7.4

Imidazole (50 mmol/L)	3.4 g
Sodium chloride (NaCl) (100 mmol/L)	5.85 g
Water to make volume up to 1 L	-

Dissolve in approximately 500 mL of water. Add 18.6 mL of 1mol/L HCl to make up the volume to 1 L with water. Store at room temperature (18–25°C).

Phosphate Buffer, Iso-osmotic

Stock solutions	
(A) KH_2PO_4 (150 mmol/L)	23.4 g/L
(B) Na_2HPO_4 (150 mmol/L)	21.3 g/L

PH	**$NaH_2PO_4 . 2H_2O$ (150 mmol/L)**	**Na_2HPO_4 (150 mmol/L)**
5.8	87 mL	13 mL
6.0	83 mL	17 mL
6.2	75 mL	25 mL
6.4	66 mL	34 mL
6.6	56 mL	44 mL
6.8	46 mL	54 mL
7	32 mL	68 mL
7.2	24 mL	76 mL
7.4	18 mL	82 mL
7.6	13 mL	87 mL
7.7	9.5 mL	90.5 mL

Phosphate Buffered Saline

Equal volumes of iso-osmotic phosphate buffer and 9 g/L NaCl.

Phosphate Buffer, Sorensen's

Stock solutions	66 mmol/L	100 mmol/L	150 mmol
(A) KH_2PO_4	9.1 g/L	13.8 g/L	20.7 g/L
(B) Na_2HPO_4	9.5 g/L	14.4 g/L	21.6 g/L
or Na_2HPO_4. $2H_2O$	11.9 g/L	18 g/L	27.1 g/L

To obtain a solution of the required pH, add A and B in the indicated proportions.

PH	Solution A	Solution B
5.4	97	3
5.6	95	5
5.8	92.2	7.8
6 k	88	12
6.2	81	19
6.4	73	27
6.6	63	37
6.8	50.8	49.2
7	38.9	61.1
7.2	28	72
7.4	19.2	80.8
7.6	13	87
7.8	8.5	91.5
8	5.5	94.5

(KH_2PO_4: Potassium dihydrogen phosphate; Na_2HPO_4: Disodium hydrogen phosphate)

Tris–Hydrochloric Acid Buffer (200 mmol/L)

Tris (hydroxymethyl) aminomethane (24.23 mmol/L)	250 mL

To obtain a solution of the required PH, add the appropriate volume of 1 mol/L HCL and then make up the volume to 1 L with water.

PH	Volume (mL)	PH	Volume (mL)
7.2	44.5	8.2	23
7.4	42	8.4	17.5
7.6	39	8.6	13
7.8	33.5	8.8	9
8	28	9	5

100 mmol/L, 150 mmol/L, 300 mmol/L, and 750 mmol/L stock solutions may be similarly prepared with appropriate weights of Tris and volume of acid.

Tris–Hydrochloric Acid Bovine Serum Albumin Buffer, pH 7.6 (20 mmol/L)

Tris (hydroxymethyl) aminomethane (20 mmol/L)	2.42 g
EDTA, disodium salt (10 mmol/L)	3.72 g
NaCl (100 mmol/L)	5.85 g
Sodium azide (3 mmol/L)	0.2 g
HCL (10 mmol/L)	–
Bovine serum albumin	10 g
Water to make volume up to 1 L	–

Dissolve the reagents in 800 mL of water. Adjust the pH to 7.6 with 10 mol/HCl. Add 10 g of bovine serum albumin (BSA) and make up the volume to 1 L with water.

Role of buffer in electrophoresis

Buffer ions have a two-fold purpose in electrophoresis: carry the applied current and fix the pH at which electrophoresis is carried out and, thus, determine the kind of electrical charge on the solute, and the extent of ionization of the solute. By these effects, the buffer determines the direction of the electrophoretic migration to one of the two electrodes. The buffer ionic strength determines the thickness of the ionic cloud surrounding the charged molecules and, thus, also the rate of migration and the sharpness of the zones.

Many buffer systems are used in electrophoretic procedures, but the most widely used are the barbitone buffers and the Tris–Boric acid–EDTA buffers.

Viva Questions

1. What is Henderson–Hasselbalch equation and its application?
2. How does a pH meter function? How is pH measured?
3. What are buffers?
4. Explain briefly the preparation of buffers.

25 Biomedical Waste Management

BI 11.1

Biomedical waste (BMW) is the waste that is generated in hospitals and laboratories during diagnosis, treatment and research activities or from immunisation of human beings or animals. Of the total quantity of waste generated by healthcare activities, almost 80–90% is general waste, comparable to domestic waste. This comes from the administrative and housekeeping activities of hospitals and laboratories. The remaining 10–20% of waste is considered hazardous and infectious.

Biomedical waste management is done in 4 steps.

Step – 1: Segregation

Step – 2: Collection and storage

Step – 3: Transportation

Step – 4: Treatment and disposal

Segregation of Waste

Containers that carry BMW are made of specific colours for easy identification of the different types of waste (Table 25.1). All containers with different coloured plastic bags/containers should be located at the point of generation of the waste, e.g., labs, medical and surgical wards and operation theatres. It should be ensured that waste bags are filled only to three-fourth capacity, tied securely and removed from the site of the generation regularly and promptly. Certain categories of waste such as plastics and sharp materials may need prior treatment at the site of generation (decontamination/disinfection) and should be removed from the site of generation only after treatment. Biomedical wastes are classified into ten categories according to their treatment and disposal, taking into account their characteristics (Table 25.1).

Storage of Waste

Untreated BMW should not be stored beyond 48 hours. It should be ensured that waste does not adversely affect human health and environment.

Transportation of Waste

Waste routes should be designated and a specific time should be earmarked for BMW transportation to limit the chances of its mixing with general waste.

Separate wheeled trolleys should be used and they should be cleaned and disinfected in case of any spillage. Trolleys should not contain any sharp edges and should be easy to clean.

BMW should be transported only in vehicles authorised by the competent authority, as specified by government.

Treatment and Disposal of Waste

General Waste

About 80–90% of the waste generated in hospitals is general waste. This waste is non-toxic and non-infectious. It usually comprises paper, leftover food articles, peels of fruits, disposable paper glasses for tea/coffee, cardboard boxes, wrappings or outer covers, etc.

Such general waste should be put into black coloured polyethylene bags and deposited at the municipal dumps to be subsequently collected by local civic authorities.

Biomedical Waste (BMW)

The BMW must be disinfected and mutilated/shredded before its final disposal. Anatomical waste must be buried deep. Syringes should be cut using hub cutters and chemically disinfected with 1% bleaching powder solution at the source of generation before final disposal into pits meant for sharps (sharp pits). Infected plastics should be chemically disinfected or autoclaved, shredded and recycled and sent for final disposal to municipal dumps.

The treatment options for BMW are discussed here briefly. A comprehensive list of global rules for BMW disposal was framed in 1998, which is updated periodically.

Incineration

The combustion efficiency and emission levels of the incinerator must be as per the BMW disposal rules 1998. A suitably designed pollution control device certified by the pollution control board must be installed. In case of small hospitals/labs, a common incinerator can be jointly used. Waste to be incinerated should not be chemically treated with chlorinated disinfectants. Chlorinated plastic bags should not be incinerated. The ash produced by the incinerator should be sent for land-filling; it should also be periodically checked for toxic metals.

Deep burial

The waste under categories 1 and 2 (see column 3 of Table 25.1) should be subjected to deep burial at locations away from residential areas. It must be ensured that no contamination of any surface or ground water occurs, and that the location is not prone to flooding or erosion.

Autoclave and microwave

BMW categories 3, 4, 6 and 7 (see column 3 of Table 25.1) must be treated by these techniques.

Shredding and needle destruction

Plastics (IV bottles, syringes, catheters, etc.) and sharps (needles, blades, scalpel, etc.) should be shredded after chemical treatment/microwaving/autoclaving. Needle destroyers are used for disposal of needles directly, without chemical treatment.

Secured landfill

Incinerator ash, discarded medicines, cytotoxic substances and solid chemical waste should be treated by this method.

Sharp pit

Sharp waste should be disposed in a pit (2–5 m deep and 1–2 m wide) after disinfection. After digging, the pit should be lined with bricks and concrete rings. The pit should be covered with heavy concrete slab in which a galvanized steel pipe is arranged, projecting about 1.5 m above the slab with an internal diameter of up to 20 mm. When the pit is full, it can be sealed completely after another pit has been prepared. It should be half-filled with waste and then covered with lime up to 50 cm of the surface before filling the rest of the pit with soil.

Safety Measures

Healthcare workers must use personal protection equipment while handling BMW. These include gloves, masks, protective glasses, plastic aprons, gum boots, etc. Healthcare workers should be immunised for hepatitis B and tetanus. All waste handling workers should be made aware of the risks involved in handling BMW.

Table 25.1 *Colour coding of bags for biomedical waste segregation and treatment as per 2016 rules*

Colour of the bag	Type of waste	Waste treatment
Yellow 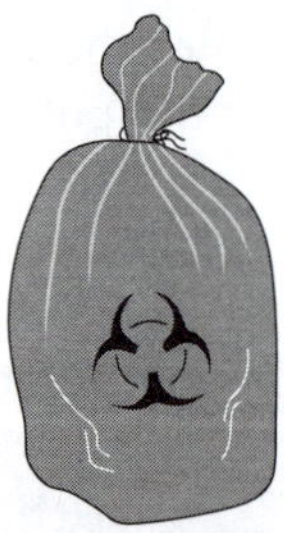	(a) Human anatomical waste	1. Incineration or plasma pyrolysis or deep burial
	(b) Animal anatomical waste	
	(c) Soiled waste	2. Incineration or plasma pyrolysis or deep burial
	(d) Expired or discarded medicines	3. Returned to the manufacturer or supplier for incineration at temperature >1200°C
	(e) Chemical waste	4. Incineration or plasma pyrolysis or deep burial or encapsulation
	(f) Chemical liquid waste	5. Pre-treatment followed by disposal
	(g) Discarded linen, mattresses and bedding contaminated with blood or body fluids	6. Non-chlorinated chemical disinfection followed by incineration or plasma pyrolysis
	(h) Microbiology, biotechnology, and other clinical laboratory waste	7. Pre-treated to sterilise with non-chlorinated chemicals on-site as per NACO or WHO guidelines; thereafter, sent for incineration
Red 	Contaminated waste (recyclable) like plastic bag, bottles, pipes or container	8. Autoclaving or microwaving/hydroclaving, followed by shredding or mutilation. Treated waste to be sent to registered recyclers for energy recovery, or for recovery of plastics, diesel or fuel oil or for use in road-making

Colour of the bag	Type of waste	Waste treatment
White translucent puncture-proof container 	Sharp waste including metals: needles, syringes with fixed needles, needles from needle tip cutter or burner, scalpels, blades	9. Autoclaving or dry-heat sterilisation, followed by shredding or mutilation or encapsulation in metal container or cement concrete. Send for final disposal to iron foundries (having consent to operate from the state pollution control committees) or sanitary landfill or designated concrete sharp waste pits
Blue cardboard box with blue label or blue leak- and puncture-proof container 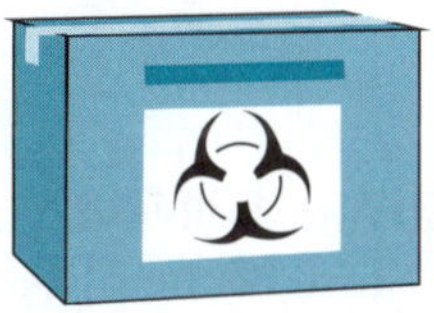	Glassware: Broken or discarded and contaminated glass including medicine vials and ampoules except those contaminated with cytotoxic wastes; metallic body implants	10. Disinfection (by soaking the washed glass waste after cleaning with detergent and sodium hypochlorite treatment) or through autoclaving or microwaving or hydroclaving; then sent for recycling

Viva Questions

1. What is biomedical waste? What are the steps of biomedical waste management?
2. What are the different types of biomedical wastes?
3. How is biomedical waste treated and disposed off?

26 DNA Isolation and Quantification

BI 11.16

DNA is the genetic repository of all the living forms. Understanding the biological processes at gene level is currently a rapidly developing field in molecular biology. Most of the deviations in biological processes have a basis at genetic level that ultimately results in a disease. So, gene manipulation techniques including gene therapy is the ultimate goal of all research on DNA.

Extraction of DNA and its estimation is the first step in any investigation on DNA. Hence, it is important to be familiar with this basic step. Common DNA extraction methods include organic extraction, Chelex extraction and solid phase extraction.

Basic Steps of DNA Isolation

In order to study DNA, it has to be first extracted from the cell. In eukaryotic cells such as human and plant cells, DNA is organised as chromosomes in the nucleus. Since bacterial cells do not have nucleus, DNA in them is organized in rings or circular plasmids in cytoplasm. The DNA extraction process frees DNA from the cell and then separates it from the cytoplasm and proteins.

The basic steps of DNA extraction are (i) lysis, (ii) precipitation and (iii) purification.

1. **Lysis:** In this first step of DNA extraction, DNA is released from the cell and nucleus in two ways. First, by mechanical disruption the cells are broken up. This is achieved with a tissue homogenizer (using a small blender or mortar and pestle) or by cutting up the tissues into small pieces. Mechanical disruption is used to open plant cells because they have a rigid cell wall. Second, lysis is achieved with detergents and enzymes such as proteinase K to free the DNA by dissolving the associated proteins.
2. **Precipitation:** After the first step the DNA has been freed from the nucleus but is still mixed with cellular debris. Precipitation is used to separate out the debris. First, sodium ions (Na^+) are used to neutralise the negative charged groups on the DNA molecules to make them more stable and water soluble. This is followed by treatment with alcohol (such as ethanol or isopropanol) to precipitate DNA out of the aqueous solution.
3. **Purification:** After precipitation in the aqueous phase, the DNA is rinsed with alcohol once again to remove remaining unwanted remnants of the cellular debris. At this stage, the DNA is almost purified and is re-dissolved in water for easy handling and storage.

Why is DNA purity important?

Accurate measurement of DNA concentration is critical for many applications in molecular biology. Impurities in DNA lead to inaccurate measurements of DNA concentration, which adversely affect sensitive techniques.

Isolation of DNA by Phenol–Chloroform Method

Firstly, the cells are lysed using a detergent and then digested with a specific enzyme, proteinase. The proteins are then separated from the cell by subjecting them to repeated extraction with phenol and chloroform. Proteins being soluble in phenol/chloroform, the interphase between the phenol and aqueous layer is separated out after centrifugation while DNA and RNA are trapped in the aqueous layer. RNA is removed by RNase treatment and DNA is precipitated out using ethanol and sodium acetate.

The DNA concentration is measured using a double cell UV spectrophotometer. The nucleic acids absorb light strongly in the UV region at 260 nm due to conjugated bonds present in the purines and pyrimidine basis.

Reagents used

i. Tris–EDTA (TE) buffer (pH 8.0) containing 10 mM Tris, 1 mM EDTA
ii. DNA lysis buffer – 3% SDS (sodium dodecyl sulfate) in 2X TE buffer (pH 8.1)
iii. Proteinase-K – Stock 20 mg/mL in 20 mm Tris HCl buffer (pH 7.5)
iv. Sodium acetate – 3M
v. Saturated phenol in TE – Contains 2 mg of 8-hydroxy-quinoline in 100 mL TE buffer.
vi. Chloroform: Isoamyl alcohol (CIA) (24:1)

Procedure

Take a small piece of tissue from the biopsy sample or scrapping and mince in a mortar and pestle using 0.5 mL of 1X TE buffer and a pinch of sand. Add 2 mL of 1X TE buffer and transfer the tissue homogenate to a 15 mL falcon tube. Rinse mortar with further 1.5 mL of TE buffer. Add lysis buffer in ratio of 1:2, i.e., approximately 50% of the total volume of the homogenate. Add such a volume of proteinase K (10 mg/mL) in the final volume of the solution. Incubate at 37^0C overnight.

Then add equal volume (6 $\simeq$ mL) of phenol and mix thoroughly in shaker for 15 minutes. Centrifuge at 5000 rpm for 2 minutes at 4^0C. Take the supernatant in a fresh tube using Pasteur pipette. Then add an equal volume of phenol and CIA (1:1). Shake in shaker for 15 to 30 minutes. Centrifuge and take the supernatant in a fresh tube. Repeat the step. Then to the supernatant add equal volume of CIA. Shake for 15 minutes and then centrifuge at 5000 rpm for 10 minutes at 4°C. To the aqueous supernatant add 1/10th volume of chilled 3M sodium acetate solution and 2.5 volume of chilled absolute ethanol. DNA forms a precipitate and can be seen as small coils that can be spooled out in a tube or pelleted by centrifugation. Wash the DNA pellet with 70% ethanol and air dry. Add 100–200 µL of 1X TE buffer to dissolve the pellet.

Store the DNA at –20°C to – 70°C till use.

Quantification of DNA

Quantification of DNA is important because it is necessary to know the quantity of DNA that is present when performing techniques such as PCR, RFLP and RAPD.

i. Make an accurate dilution of the DNA in distilled water and mix well.
ii. Fill three quarters of the reference cell of spectrophotometer with distilled water plus some TE buffer. Avoid cross contamination and bubble formation in cells. Similarly, fill the sample cell with diluted DNA.

iii. Adjust spectrophotometer sequentially to measure optical density (OD) at 260 nm and 280 nm.
iv. Note the OD values and calculate the purity and concentration of DNA.

Precautions

i. Stress should be avoided to prevent shearing of DNA. So no vortex mixing or vigorous pipetting should be done.
ii. All glassware, plasticware and reagents must be autoclaved before use.
iii. Phenol is corrosive, so it should be handled carefully.

Observations and calculation

OD at 260 nm = …….
OD at 280 nm = ……...

(Nucleic acids have an absorbance wavelength of 260 nm because of the nucleobases that they are made of. Proteins, on the other hand, tend to absorb light at 280 nm.)

Quantification

It is done by taking the OD value equal to 1.0 of a DNA sample of concentration 50 µg/mL at 260 nm, path length of 1 cm. Hence the DNA concentration in the sample can be calculated by taking the OD value of the DNA sample and dilution factor into consideration. It is usually expressed as µg or ng/mL of sample.

Absorbance at 280 nm is due to proteins; at 260 nm it is due to DNA. Hence the ratio absorbance at 260 nm to the absorbance at 280 nm gives indication about the purity of DNA sample. This ratio for a pure DNA sample is 1.8 (range of 1.7 to 2.0).

Viva Questions

1. State the importance of DNA isolation and quantification.
2. How is DNA isolation achieved?
3. How is DNA quantification done?

27 Biochemical Tests for Common Clinical Conditions

BI 11.17

In this chapter the basis and rationale of the biochemical tests carried out for common clinical conditions listed below are briefly explained.

1. Diabetes mellitus
2. Dyslipidemia
3. Myocardial infarction
4. Renal Failure
5. Gout
6. Proteinuria
7. Nephrotic syndrome
8. Edema
9. Jaundice
10. Liver diseases
11. Pancreatitis
12. Acid–base disorders
13. Thyroid disorders

1. Diabetes Mellitus

Diabetes mellitus (DM) is one of the worlds most significant public health problems. Diabetes is a common clinical condition affecting above 10% of the general population (more prevalent in those above the age of 50). It is a metabolic disorder that is caused either by the **deficiency of insulin** or by the body's **resistance to its action,** resulting in increased blood glucose levels **(hyperglycemia),** which lead to several systemic complications.

Diabetes mellitus is broadly **classified into two categories** based on etiology rather than treatment.

1. Type 1 diabetes mellitus
2. Type 2 diabetes mellitus

Type 1 Diabetes Mellitus

Based on its treatment, this disorder was formerly known as **insulin-dependent diabetes mellitus (IDDM).** About 5–10 percent of all cases of diabetes mellitus belong to this category. The disease is caused by a loss of pancreatic β-cell function, resulting in the deficiency of insulin. The patients are dependent on insulin injections (hence, the name IDDM) to maintain normal metabolism. The onset of the disease is usually abrupt, usually in childhood and adolescence (for this reason, it is also called **juvenile-onset diabetes**—the majority are affected before the age of 30). Type 1 DM is considered to be an **autoimmune disease** in which autoreactive T-cells of the immune system destroy the β-cells of the pancreatic islets. However, the α, β and other islet cells are preserved. It is estimated that the autoimmune process leading to type 1 DM begins at least 10 years before the onset of the clinical presentation and that about 80–90 percent of β-cells must be destroyed to induce the symptoms of the disease. Hence, the early detection of these autoantibodies in persons belonging to the high-risk groups has much diagnostic value. Patients of type 1 DM are more prone to developing ketoacidosis, which has to be treated as a medical emergency.

Type 1 DM is subclassified as type 1a (immune-mediated—where autoantibodies are found in the patients) and type 1b (idiopathic—patients who present no evidence of autoimmunity).

A complex multigenic trait seems to be the mode of inheritance of type 1 diabetes.

Type 2 Diabetes Mellitus

About 90% of diabetic patients belong to this category. Unlike type 1 DM, this type of diabetes usually affects an individual after the age of 40 years (hence, it is also called **adult-onset diabetes**). Such patients are not prone to ketosis and are not dependent on insulin (it was formerly known as **non-insulin dependent diabetes mellitus—NIDDM).** Unlike type 1 DM, there is no deficiency of insulin in this condition, but its action is impaired **(insulin resistance)**. β-cell dysfunction resulting in defective insulin secretion is only secondary to insulin resistance. **Insulin resistance is defined as a decreased biological response to normal levels of circulating insulin.**

The molecular defects resulting in insulin resistance and defective insulin secretion are caused by a combination of environmental and genetic factors. Type 2 DM is most commonly associated with obesity. Dietary manipulation, weight loss and oral hypoglycemic agents improve hyperglycemia. The genetic factors contributing to the development of type 2 diabetes have not been fully identified. The genes involved in insulin secretion, its action and those regulating body weight are likely to be responsible.

Laboratory diagnosis and monitoring of diabetes mellitus

Laboratory diagnosis of diabetes is commonly done by estimating plasma glucose and the corresponding levels of glucose in urine. Investigations associated with the initial diagnosis and monitoring of DM are briefly described below.

(i) Blood glucose estimation

Blood sugar estimated at any time of the day without any prior preparations is known as **random blood sugar (RBS).** Sugar estimated in the morning, before breakfast, after an overnight fast of 10–14 hours is called **fasting blood sugar (FBS).** If the analysis is done two hours after a meal, it is called **post-lunch** or **post-prandial blood sugar (PLBS** or **PPBS)**.

For the initial diagnosis or monitoring of the glycemic status of a diabetic patient, venous blood samples are tested for glucose levels in the fasting and post-prandial states (see also Chapter 5). The whole blood concentration of glucose is approximately 10% lower than plasma concentration because RBC contains relatively lesser glucose. Plasma glucose concentration has to be maintained within normal limits because persistent hyperglycemia causes several complications.

(ii) Urine testing for glucose, ketones and proteins

The testing of **urine for glucose** is commonly done in medical wards using glucose-specific dipsticks, a method based on the principle of **Benedict's test**. Urine that was passed one to two hours after a meal is preferred to a fasting sample. It should be noted that a positive Benedict's test (glycosuria), by itself, is not diagnostic of diabetes because of individual variations in the renal threshold. Further, glycosuria is common during pregnancy. However, glycosuria must always be investigated further by testing blood for glucose levels both in healthy individuals and in pregnant women (to rule out gestational diabetes mellitus). For more details on glycosuria, refer to 69.

Ketone bodies are detected in urine by the **Rothera test**. Ketonuria is observed in healthy people who indulge in frequent fasting, those who engage in strenuous exercise for long periods and those who have been on a high-fat and low-carbohydrate diet for some time. When ketonuria is associated with glycosuria, it indicates diabetes mellitus that has been left untreated.

Testing for **urinary albumin** is performed to assess renal damage. **Excretion of more than 300 mg/day of albumin is diagnostic of diabetic nephropathy.** When albumin is excreted in levels **between 30 mg/day** and **300 mg/day**, the condition is called **microalbuminuria,** which indicates that the person is at risk of developing diabetic nephropathy. Diabetic patients must be checked for microalbuminuria at least once a year.

(iii) Blood lipids and other parameters

The concentrations of serum lipids—**total cholesterol, LDL** and **HDL cholesterol** and **triglycerides** (see Chapter 12)—are gauged to assess the metabolic status of diabetic patients.

The estimation of **uric acid, homocysteine** and **lipoprotein (a)** in blood gives valuable information about cardiovascular risk in diabetics.

(iv) Glycated hemoglobin

Diabetic patients on an oral hypoglycemic drug or on insulin therapy are advised to undergo periodic monitoring of their glycemic status (glucose concentration in blood) to assess the efficacy of treatment. Non-enzymatic addition of a sugar to a protein is called *glycation*; this occurs as a post-translational modification. In persistent hyperglycemia, proteins undergo extensive glycation. Among the glycated proteins, hemoglobin provides the most reliable information regarding hyperglycemia. Glycation usually occurs on the HbA fraction of hemoglobin. HbAlc constitutes most of the HbA1 fraction. **The concentration of HbAlc reflects the mean blood glucose concentration of the preceding 8-10 weeks.** This corresponds to the lifespan of RBC (120 days)—half of the erythrocytes are replaced in 60 days. Hence, a single estimation of HbAlc gives quite a long-term picture (of about 2 months) of glycemic status. When the level of HbAlc is less than 7% (equal to 154 mg/dL), the diabetic patient is considered to be maintaining adequate control. Above 9% (corresponding to 212 mg/dL) indicates poor control. When the value of HbA1c is between 5.7 and 6.5, the person can be diagnosed as being at a stage of impaired glucose tolerance (**pre-diabetes**). See Table 27.1 for reference values of HbA1c.

Table 27.1 *Reference values of HbA1c*

HbA1c	Glucose mg/dL)
5%	97
6%	126
7%	154
8%	184
9%	212
<5.7	Normal
>6.5	Diabetes

Table 27.2 *Biochemical parameters evaluated in DM*

	Biochemical test	Rationale of the test
1.	Increased glucose in plasma and urine	Hyperglycaemia results from improper insulin secretion, action, or both. Glucosuria occurs when renal tubular transporter system for glucose becomes saturated. This happens when plasma glucose concentration exceeds roughly 180 mg/dL in an individual with normal renal function and urine output.
3.	Glycated hemoglobin (HbA1c)	In persistent hyperglycemia, glycation of proteins usually occurs on the HbA fraction of hemoglobin. HbAlc constitutes most of the HbA1 fraction. The concentration of HbAlc reflects the mean blood glucose concentration of the preceding 8–10 weeks. This corresponds to the lifespan of RBC (120 days).
2.	Increased urine specific gravity	Glucosuria increases the solute content and ultimately the specific gravity of urine.
3.	Serum insulin and C-peptide	**In type 1 diabetes,** absence of insulin and decreased C-peptide in the blood. As C-peptide is produced along with insulin from proinsulin, its levels are also decreased with insulin due to β cell failure. **In type 2 diabetes**, due to peripheral resistance of cells to insulin, the plasma levels increase and finally decrease in the later stages of the disease.
4.	Increased serum and urine osmolality	Serum osmolality is high as a result of hyperglycaemia.
5.	Ketones in serum and urine (in ketonemia and ketonuria)	Uncontrolled diabetes leads to β oxidation of fatty acids in adipose tissues, which form acetone, acetoacetate and β hydroxyl butyrate (ketone bodies) in liver. This is released in the blood and ultimately, in urine when in excess.

	Biochemical test	Rationale of the test
6.	Decreased blood and urine pH (in acidosis)	• Acetoacetate and β hydroxyl butyrate contribute to acidosis mainly. • Lactate, fatty acids and other organic acids can also contribute to a lesser degree. • Bicarbonate and total CO_2 are usually decreased due to Kussmaul breathing.
7.	Electrolyte imbalance	• Sodium concentration tends to be low due to loss (polyuria) and in a part due to a shift of water from the cells because of the hyperglycaemia. • Hyperkalemia is almost always present as a result of displacement of potassium from cells in acidosis.
8.	Impaired lipid profile	• Triglycerides are increased in blood. • Fatty acid oxidation is inhibited in type 2 diabetes. This causes fatty acids to be incorporated in glycerol as triglycerides (triacylglycerol) for the release of very low-density lipoproteins in blood.

2. Dyslipidemia

Diseases associated with abnormal lipid concentrations are referred to as dyslipidemias. They may be familial; can be caused directly by genetic abnormalities or through environmental/lifestyle imbalances, or they can develop secondarily as a consequence of other diseases. Dyslipidemias are generally identified by the clinical characteristics of patients and the results of laboratory tests.

Cholesterol

Elevated levels of lipids in plasma, especially cholesterol, are of tremendous clinical significance. They are associated with the development of **atherosclerosis,** a phenomenon underlying majority of **cardiovascular diseases.** Cardiovascular disease is one of the most common causes of mortality world over.

Plasma cholesterol

In healthy persons, the levels of total plasma cholesterol vary between **150–200 mg/dL.** However, it should be *preferably below 180 mg/dL.* Women of reproductive age have relatively lower levels of cholesterol, which is attributed to the female sex hormones. In the newborn, the plasma cholesterol is about 100 mg/dL and rises to about 150 mg/dL in the first year of life. Plasma cholesterol levels start rising after the fourth decade of life in men and after menopause in women.

Principle of estimation

Estimation of total cholesterol and its lipoprotein fractions has tremendous clinical significance in the assessment of cardiovascular risk. HDL-cholesterol can be directly measured.

Unlike HDL-C which is directly estimated, LDL-C is indirectly calculated from the serum triacylglycerol (TAG) level because the separation of LDL from the serum is a cumbersome process.

A formula **(Friedewald equation)** is usually employed to estimate LDL-C levels. According to the Friedewald equation, the total cholesterol (Tc) is equal to the sum of cholesterol present in HDL-C, LDL-C and VLDL-C.

$$TC = HDL\text{-}C + LDL\text{-}C + VLDL\text{-}C \text{ (all in mg \%)}$$

$$LDL\text{-}C = TC - [HDL\text{-}C + VLDL\text{-}C]$$

$$= TC - [HDL\text{-}C + TAG/5]$$

Usually, VLDL-C is taken as 1/5th of the serum triacylglycerol (TAG) level, based on the assumption that the ratio of TAG and cholesterol content of VLDL is 5:1. However, this calculation is not applicable when the TAG level exceeds 400 mg/dL or if the patient has type III hyperlipoproteinemia, since these conditions alter the usual composition of VLDL.

HDL-C and LDL-C ratio

The ratio of LDL-C:HDL-C is considered a valuable marker to assess the risk of atherosclerosis, better than the individual levels of LDL-C or HDL-C. Normal LDL-C:HDL-C is 2–2.5. A high ratio is a predictor of risk of CAD.

In healthy adults, the LDL-cholesterol is about 80–130 mg/dL, while HDL-cholesterol is in the range of 30–70 mg/dL.

Cholesterol and heart disease

From the above discussion it is conclusive that low levels of LDL-cholesterol and high levels of HDL-cholesterol are desirable. Elevated total plasma cholesterol levels (>240 mg/dL) and LDL-cholesterol levels (>160 mg/dL) are implicated in the development of **atherosclerosis,** which leads to **coronary artery disease (CAD)**. Hypercholesterolemia causes the deposition of cholesterol and cholesteryl ester in the connective tissue of the arterial walls, resulting in blockage of coronary arteries and cerebral blood vessels.

Laboratory analytes of lipid metabolism and cardiovascular disease

Elevated levels of lipids are always viewed as a warning signal for cardiovascular disease. Elevated levels of serum cholesterol and other lipid fractions are associated with the incidence of atherosclerosis and **coronary heart disease (CHD)**.

Plasma lipid profile

Measurement of serum cholesterol alone is not sufficient to assess the risk of cardiovascular disease. Other lipid parameters in plasma are equally important. **Plasma lipid profile** comprise the relative concentrations of the following lipid fractions in plasma:

1. Total cholesterol (TC) (140–200 mg/dL)
2. Triacylglycerols or triglycerides (TAG or TG) (60–150 mg/dL)
3. HDL-cholesterol (HDL-C) (40–75 mg/dL)
4. LDL-cholesterol (LDL-C) (60–130 mg/dL)

To estimate the plasma lipid profile, a sample of serum should be taken after 12 hours of fasting.

Elevated levels of TC and LDL-C are potential risk factors for **coronary artery disease (CAD)**. However, elevated HDL-C offers protection against this condition. Elevated levels of TAG are less important as a risk factor for CHD, but they can cause **pancreatitis.**

Apart from the plasma lipid profile, in some cases the following plasma estimations are also needed to study the risk of cardiovascular disease.

1. ApoB level
2. ApoA level
3. Lp(a) level
4. 'Small dense' LDL level
5. Homocysteine level
6. High-sensitivity C reactive protein (hs-CRP) level

Risk factors for cardiovascular disease

Elevated levels of plasma lipid profile and other associated lipid fractions are indicative of a definite risk of cardiovascular disease. Obesity and a sedentary lifestyle, cigarette smoking, diabetes mellitus and hypertension, if not under control, contribute to coronary heart disease.

Atherosclerosis

When the levels of LDL-C are elevated, LDL is oxidised. **Oxidised LDL** particles are deposited on the walls of the arteries. Macrophages take up the LDL-C and become **foam cells.** These changes lead to the formation of atherosclerotic **plaque** that causes the narrowing of the lumen of the vessel. The turbulent flow of blood through these occluded arteries favours clot formation. **Thrombosis** of the coronary and cerebral vessels causes **ischemia** (lack of blood supply) to the vital organs. This leads to **coronary artery disease (CAD)** and **stroke.** Sometimes, an occlusive thrombus in a coronary artery leads to **myocardial infarction (MI),** which could be fatal.

3. Myocardial Infarction

Myocardial infarction, commonly called heart attack, occurs when blood flow decreases or stops to a part of the heart, causing damage to the heart muscle. The common cause for MI is coronary artery disease (CAD) that is left untreated. Hypertension, chronic heavy smoking, elevated lipid levels in the blood, etc., are risk factors for MI. The most common symptom is chest pain or discomfort, which may radiate to the shoulder, arm, neck, jaw or back.

In myocardial infarction (MI), a clot is formed which occludes one of the major vessels. Thrombosis in coronary artery leads to ischemia of cardiac tissue due to the hindrance in oxygen supply, finally leading to infarction (death of tissue).

After an occurrence of MI, a number of intracellular enzymes and proteins are released from the damaged cells. They are of diagnostic importance and are called **cardiac markers.** These biochemical markers are categorised as:

a. Cardiac markers for myocardial infarction
b. Cardiac markers for risk prediction
c. Cardiac markers for myocardial stress

Cardiac Markers for Myocardial Infarction

	Marker	Mechanism	Outcome
1.	CK-MB (creatinine kinase-MB)	CK-MB has been shown to be most specific for myocardial cells.	CK-MB rises within 4–6 hours after the onset of chest pain, peaks at 12–24 hours and returns to normal levels within 2–3 days.
2.	Troponins (T and I)	The major functions of troponin is to bind calcium and regulate muscle contractions. Following injuryto cardiac muscle cells, the troponin complex and free troponin subunits are released into the blood stream.	Troponins rise 4–10 hours after the onset of symptoms, peak at 12–48 hours and remain elevated for 4–10 days.
3.	Myoglobin	Myoglobin is an oxygen-binding heme protein present in both cardiac and skeletal muscles. It lacks specificity.	Myoglobin rises as early as 1–4 hrs after the onset of symptoms and rapidly returns to baseline within 18–24 hours.
4.	LDH and AST	The use of LDH and AST is declining rapidly because of non-specificity.	

Cardiac Markers for Risk Prediction

1. Lipid profile
2. HbA1c levels
3. Serum homocysteine

Cardiac Marker for Myocardial Stress

Brain natriuretic peptide (BNP) is present in the human brain but more in the cardiac ventricles. The ventricles secrete BNP in response to excessive stretching of the heart muscle cells. Patients with congestive cardiac failure have increased plasma concentration of BNP.

Clinical case study

A 60-year-old man was brought to hospital with complaints of severe chest pain, breathlessness and vomiting. He could be brought to the nearby referral hospital in town only 5 hours after the onset of chest pain. His blood was immediately drawn at the casualty and investigations were carried out. The results are as follows:

Investigation (in serum)	**Result**	**Reference range**
Creatine kinase (CK)	400 IU/L	15–100 IU/L
Asparatate transaminase (AST)	70 IU/L	8–20 IU/L
Lactate dehydrogenase (LDH)	380 IU/L	100–200 IU/L

It is noted that the levels of CK-MB and LDH-1 are in particular highly elevated.

Diagnosis: It is a clear case of **myocardial infarction (MI)**. The man suffered heart attack as is evident from the elevated levels of CK-MB and LDH-1 and AST. The CK level starts to rise within 3–6 hours of the infarction. Hence, CK estimation is very crucial to the early diagnosis of MI. CK levels reach a peak within 18–24 hours after an attack. The elevations in LDH and AST further confirm the diagnosis. Thrombolytic therapy must be started immediately to save the life of the patient.

Nowadays, apart from the above 3 enzymes, the estimations of myoglobin and cardiac troponins are also being carried out to prove early diagnosis of MI.

The MI injured heart muscle releases CK–MB and other enzymes into blood. Hence any elevation of these enzymes (known as cardiac markers for this reason) are diagnostic of MI.

4. Renal Failure

Kidney failure occurs when your kidneys lose the ability to sufficiently filter waste from your blood and maintain body fluid balance. The term 'renal failure' covers many problems. Long-standing disease of the kidneys leads to renal failure (**chronic renal disease**). When the kidneys suddenly fail to filter waste from blood, it is called **acute renal failure**.

Kidneys can get damaged from a physical injury or long-term diseases such as diabetes and hypertension. Both these diseases, when not treated, are potential risk factors for kidney failure. The condition does not usually occur all of a sudden; it is the end result of a gradual loss of kidney function. When the kidneys lose their filtering capacity, abnormal levels of waste products accumulate in the blood, severely impairing the composition of body fluids. Initially, kidney failure may not cause any symptom.

Common symptoms of kidney failure such as oliguria (reduced amount of urine), pedal edema (swelling in the legs and feet) occur as early signs because of the retention of fluids due to the failure of kidneys to eliminate waste water. Other symptoms (signs) include shortness of breath, fatigue, confusion and abnormal heart rhythms (due to the inability to remove potassium from the blood).

Since there are multiple causes that lead to kidney failure, treating the underlying disease on time corrects the kidney abnormality. The diagnosis of kidney failure is usually made with blood tests that measure creatinine, BUN (blood urea nitrogen) and GFR (glomerular filtration rate). A lifelong effort to control hypertension and diabetes is the best way to prevent chronic kidney disease and its progression to kidney failure. As age advances, kidney function gradually decreases. If kidneys fail completely, the only treatment options are dialysis or renal transplant.

The many factors that can interfere with kidney health and function are:

1. Toxic exposure to environmental pollutants or certain medications
2. Certain acute and chronic diseases
3. Severe dehydration
4. Kidney trauma

Renal failure can be classified as
a. Acute renal failure
b. Chronic renal failure

Acute renal failure (ARF) is a clinical syndrome that is characterised by a rapid deterioration of kidney function that disorganises metabolic, electrolyte and fluid homeostasis over a period of hours to days. AKI complicates the course and worsens the outcome in a significant number of hospitalised patients.

Chronic renal failure/ Chronic kidney disease (CKD) is defined as persistent **kidney damage** accompanied by a reduction in the glomerular filtration rate (GFR) and the presence of albuminuria. It means that the **kidneys** are damaged and cannot filter blood the way they should. The **disease** is called **chronic** because the damage to the **kidneys** happens slowly over a long period of time. This damage can cause waste substances to build up in the body.

Acute renal failure is best differentiated from chronic renal failure by careful scrutiny of patient's history, imaging of the kidneys; small shrunken kidneys on ultrasound are a clear indication of chronicity. It develops rapidly and therefore, results in rapid electrolyte, acid–base and fluid imbalances.

Laboratory Investigations in Acute Renal Failure

	Biochemical test	Rationale
1.	Serum creatinine, blood urea	Increased serum creatinine to >200%–300% of the baseline value
2.	Serum electrolytes	In ARF, there is rapid electrolyte imbalance.
3.	Arterial blood gas analysis (ABG)	In ARF, acid–base and fluid imbalances are seen
4.	Novel urine markers in ARF 1. Cystatin-C 2. NGAL 3. Microalbumin	These help in early diagnosis of ARF.

Laboratory Investigations in Chronic Renal Failure

The following tests are carried out in addition to the tests mentioned above.

	Parameters	Rationale	Clinical manifestations
1.	Hemoglobin, vitamin B_{12}, folate and iron studies	Loss of erythropoietin production by kidneys	Severe anaemia
2.	Serum calcium and phosphate	Failure to form active form of vitamin D in kidneys	Hypocalcaemia, hyperparathyroidism
3.	Vitamin D and parathyroid hormone	Failure of ion homeostasis	Bone resorption, phosphate retention
4.	Arterial blood gas analysis (ABG)	Loss of bicarbonate, pH decreased	Acidosis
5.	Serum electrolytes	Loss of water balance function	Sodium retention, polyuria
6.	Urinary proteins	To rule out nephrotic syndrome	

Clinical case study

A 45-year-old man who used to have recurrent attacks of sore throat in the past developed rapid onset of headache, malaise, loin pain, hematuria, proteinura, hypertension, diminished urine output and peripheral edema. He was admitted to the Medicine ward for evaluation. Previous throat swab culture report showed that the sore throat he had in the past was due to streptococcal infection.

Go through the results of the investigations given below and comment.

Urine

Protein	2 g/24 hours
Microbiological examination	RBC +++; Pus cells +++

Blood

Total protein	5.5 g/dL
Albumin	3.8 g/dL
Cholesterol	160 mg/dL
Blood urea	80 mg/dL
Serum creatinine	2.4 mg/dL
ASO titer	Increased (Evidence of β hemolytic streptococci)
C3 complement	Decreased (Consumed in immune complex formation)

A probable diagnosis is **glomerulonephritis.** Streptococal infection is notorious for causing immune-mediated injury to glomerular filtration membrane, leading to the damage of GBM (glomerular basement membrane), which affects glomerular filtration, and consequently, renal function. The patient has the classic features suggestive of acute glomerulonephritis, namely sudden onset, hematuria, pyuria, edema, hypertension and oliguria.

5. Gout

Gout is an inflammatory response to the monosodium urate monohydrate crystals (MSU) crystals formed secondary to hyperuricemia. It is characterised by episodic joint pains due to deposition of monosodium urate crystals in joints and connective tissue (tophi) and urate stones in the kidneys (nephrolithiasis). Metatarsophalangeal joint of big toe is first affected. Diagnostic tests include investigation for elevated serum uric acid and tophi in joints. **Gouty arthritis** may be associated with urate crystals in joint fluid as well as with deposits of crystal **(tophi)** in tissues surrounding the joint.

Gout is conventionally described as primary and secondary.

Primary gout: It is either idiopathic or as a result of inborn errors of metabolism. Metabolic disorders that mostly occur due to inherited enzyme defects (xanthine oxidase) result in increased synthesis of purine nucleotide, leading to marked hyperuricemia and gout.

Secondary gout: Manifestations of gouty arthritis occur secondary to certain clinical conditions causing increased synthesis or decreased excretion of uric acid. Increased uric acid formation is seen in various types of malignancies, psoriasis and chronic alcoholism. Decreased uric acid disposal occurs in certain conditions like chronic renal disease, lead poisoning and pre-eclampsia.

Laboratory Investigations in Gout

	Laboratory investigation	Rationale
1.	Serum uric acid	Hyperuricemia is seen in primary gout due to enzyme defects, and in secondary gout due to malignancies as there is excessive turnover of cells leading to increased uric acid formation.
2.	Synovial fluid examination	Microscopic examination of synovial fluid for sodium urate crystals
3.	Renal function test, lipid profile, ESR and CRP	Helps in management of the disease

Clinical case study

A 50-year-old male business executive was admitted to hospital at midnight with complaints of excruciating pain in his right first metatarsophalangeal joint and inability to keep his foot on the floor. On examination, the joint was found to be swollen, red and extremely tender. The doctor on duty treated him with an anti-inflammatory drug (indomethacin) which brought some relief. The patient also gave a history of a previous attack of renal colic the previous year, which had not been investigated then. Regular business dinners and alcohol consumption was part of his lifestyle. What is your probable diagnosis and how would you manage the case?

Diagnosis: This is a **classic presentation of gout**. The case can be further investigated with the following lab test results.

Lab test	Result	Reference range
Serum uric acid	15 mg/dL	3–7 mg/dL
Urinary uric acid	3.0 g/day	0.5–0.7 g/day
Urinary pH	4.2	5.5–7.0
Blood urea	28 mg/dL	15–40 mg/dL
Serum creatinine	0.8 mg/dL	0.5–1.5 mg/dL

Given the classic presentation along with history and laboratory findings, the diagnosis of **gout** can be confirmed beyond doubt. The onset is sudden, at night, affecting a small peripheral joint (in this case, the first MTP joint of the big toe). The classic signs of inflammation (that is, redness, swelling, tenderness) were present and hyperuricemia was confirmed by laboratory findings. In this case, the past history of renal colic suggests the possibility of a renal urate stone. Further, it may be noted that gout is more common in men than women and is associated with lifestyle and obesity.

The diagnosis can be further confirmed, if needed, by demonstrating the presence of *tophi* and *monosodium urate crystals* in the synovial fluid by histopathological examination. The patient can be treated with the drug **allopurinol** along with *uricosuric* agents and NSAIDS (non-steroidal anti-inflammatory drugs), and advised to moderate his lifestyle.

6. Proteinuria

Proteinuria is defined as an increase in the amount of protein in urine. Normally, the total amount of protein excreted in urine over 24 hours is <150 mg, which is not detectable by routine tests. Urea in excess of this is referred as proteinuria. It is an indicator of leaky glomerulus, hence the first sign of glomerular injury. Most common cause of proteinuria is **nephrotic syndrome** and **diabetic nephropathy.**

Types of Proteinuria

1. **Glomerular proteinuria** – This is the most common and serious type of abnormal proteinuria. Most of the excreted protein is **albumin,** hence it is also labelled as albuminuria.
2. **Tubular proteinuria** – This happens when the tubular reabsorption mechanism is impaired and hence, low molecular-weight proteins appear in urine. This may occur alone, but is associated mostly with glomerular proteinuria.
3. **Overload proteinuria** – When low molecular-weight proteins are increased in blood, they overflow in urine, e.g., **hemoglobinuria** and **Bence Jones proteins.**
4. **Post renal proteinuria Bence Jones proteins** – This refers to proteins arising from the urinary tract below the kidneys and is usually caused by inflammation, bleeding or malignancies.

Protocol for Urinary Protein Measurement

1. Test random urine samples for protein with **dipsticks.**
2. If **positive, repeat** over a period of 1–2 weeks.
3. If two or three tests are positive, send a random urine sample for **protein–creatinine ratio** (spot protein-to-creatinine ratio).
4. If spot protein-to-creatinine ratio is > 0.2, then send for 24-hours urinary protein-to-creatinine ratio.

Laboratory investigations in proteinuria

Tests	Rationale
Dipstick – Point of care testing (POCT)	To detect proteinuria
A. Random urine protein B. Spot urine protein–creatinine ratio C. 24 hours urine protein estimation (Note: Normal urine proteins <150 mg/day; Proteinuria >150 mg/day)	The excretion of protein in urine depends on its cause. – In nephrotic syndrome, change in charge and size on glomerular basement membrane – In diabetes and hypertension, it is due to early renal damage. – Mild proteinuria in pregnancy is physiological.
Protein electrophoresis	For measuring and comparing the various fractions of plasma proteins for analysis

7. Nephrotic Syndrome

It is caused by diseases that increase the permeability across the glomerular filtration barrier.

Four of its characteristics features are:

1. Nephrotic range proteinuria
2. Hypoalbuminemia
3. Oedema
4. Hyperlipidemia

Laboratory investigations in nephrotic syndrome

	Laboratory investigations	Laboratory findings	Rationale
1.	Serum albumin	Hypoalbuminemia	Increased permeability of the glomerular basement membrane, which leads to heavy proteinuria (> 3 g/day)
2.	Lipid profile	Hyperlipidaemia	Due to compensatory increase in lipoproteins synthesis
3.	Protein electrophoresis	– Hypoproteinuria – Decrease in albumin band	– Massive albuminuria leads to increase in serum globulins. – Decrease in albumin band and increase in alpha-2-macroglobulin is observed in serum electrophoretogram.
4.	Urine analysis	Oval fat bodies	Lipiduria
5.	Vitamin-D	Hypovitaminosis-D	Vitamin D binding protein is excreted in urine, leading to hypovitaminosis-D.

Clinical case study

A 6-year-old girl was admitted to hospital with puffiness of face, weakness and hypertension for evaluation. The report of relevant investigations is given below. Interpret the same for the relevant condition.

Urine

Protein	5 g/24 hours
Microbiological examination	Granular and epithelial casts+

Blood

Total protein	5 g/dL (reference range at this age 6–8%)
Albumin	2.5 g/dL (reference range at this age 3.8–5.4%)
Cholesterol	320 mg/dL (reference range at this age 120–200%)
Blood urea	55 mg/dL
Serum creatinine	1.6 mg/dL

Provisional diagnosis: Nephrotic syndrome

The clinical and laboratory findings of edema, hypertension, hypoalbuminemia, proteinuria (heavy proteinuria is characterised by >3 g protein per day in urine) and hypercholesterolemia are suggestive of nephrotic syndrome. The characteristic findings of nephrotic syndrome can be caused by different types of renal conditions with evidence of inflammation, or without any evidence of inflammation (minimal change nephritis), or with proliferative changes.

8. Edema

Edema is an accumulation of fluid in the interstitial space that occurs as capillary filtration exceeds the limits of lymphatic drainage, producing noticeable clinical signs and symptoms. In other words, edema is the swelling caused by excess fluid trapped in tissues. It can affect any part of the body but is noticed more in legs, ankles and feet; it can affect face, hands and abdomen and the entire body as well.

Edema may be **localised** or **generalised.** Localised edema is due to lymphatic or venous obstruction. The causes of generalised edema include cardiac, hepatic and renal disorders, beriberi and severe anaemia. It can occur due causes that include the adverse effect of medication (some anti- hypertensives and analgesics); pregnancy (enlarged uterus causes pressure on blood vessels in the lower trunk of the body); or an underlying disease (more often congestive heart failure, renal failure, cirrhosis of liver, lung diseases, thyroid diseases and varicose veins). Certain allergies, burns, trauma or poor nutrition may also cause edema. Bilateral pitting edema with a protruding abdomen is characteristic of kwashiorkor (poor nutrition in children aged 1–5).

Pitting vs. Non-pitting Edema

Areas of pitting edema respond to pressure, usually from a hand or finger. For example, when you press on the skin with your finger, it will leave an indentation even after you remove your finger. Chronic pitting edema is often a sign of liver, heart, or kidney problems. It can also be a symptom of a problem with veins in the vicinity. Non-pitting edema, on the other hand does not cause any lasting indentation when pressure is applied. It is often a sign of a condition affecting the thyroid or lymphatic system.

Myxedema causes generalised non-pitting edema. Filariasis causes localised non-pitting edema. Cardiac edema is seen mostly on dependent parts, that is, over the legs in ambulatory patients and posterior aspects of thighs and sacral area in recumbent patients. Edema in kidney disease first appears over the eyelids and is more apparent when the patient wakes up.

Bilateral pitting edema with a protruding abdomen is characteristic of kwashiorkor (poor nutrition in children aged 1–5).

Biochemical Mechanism

The underlying biochemical mechanism of edema in most of the conditions is the decrease in protein (albumin) levels. By maintaining colloid osmotic pressure of plasma, albumin regulates fluids in various body compartments. In albumin deficiency (hypoalbuminemia), this function gets derailed resulting in accumulation of fluid in tissue spaces (edema). Low albumin levels are seen in conditions of protein malnutrition such as kwashiorkor, cirrhosis of liver [synthesis of albumin by liver is diminished, manifesting itself as ascites), renal failure (albumin is excreted in large amounts, commonly manifested as facial edema (puffiness of face)]. In congestive cardiac failure, increased hydrostatic pressure at the venous end causes edema. Accumulation of water lowers the sodium concentration, which in turn augments the secretion of aldosterone. This results in the retention of sodium along with water.

If pressing gently on the swollen area with finger for at least 5 seconds and then removing the finger leaves a dimple on the skin (pitting edema), it is a marked sign warranting medical intervention. In all the conditions resulting in edema, serum proteins and urinary proteins are estimated in the lab.

Laboratory investigations in edema

Laboratory investigations	Rationale
Total proteins, serum albumin, serum globulin	Due to loss of albumin, the osmotic pressure will be decreased leading to stasis of fluid in interstitial space.
Renal function test	To rule out kidney disease
Brain natriuretic peptide	To rule out congestive cardiac failure
Hemogram	To rule out anemia
Thyroid function test	To rule out myxedema

Management

Edema is treated according to the condition that is causing it. Treating the underlying disease cures edema. In addition, there are a few measures that can prevent fluid from accumulating. Lifestyle changes such as restricting salt intake, maintaining optimum weight, quitting smoking, and avoiding sitting or standing for prolonged periods without moving are advised. Placing a pillow under the legs while lying down will keep legs elevated above the level of heart and is helpful. Use of support stockings to put pressure on legs helps in keeping fluids from accumulating in legs and ankles.

Diuretics in the treatment of heart failure and hypertension: Diuretics are the drugs that stimulate water and sodium extraction, thus *increasing the urinary output*. The commonly used diuretics include frusemide, spironolactone and mannitol. Since sodium intake has an adverse impact on hypertension and congestive heart failure, a *low-salt-diet* is recommended in such patients.

If edema is caused by lung disease such as *emphysema* or *chronic bronchitis*, apart from treating these diseases, patients are advised to quit smoking if they are habituated to it.

9. Jaundice

Jaundice manifests as yellowish discolouration of the skin, sclera and mucous membrane. The laboratory findings in different types of jaundice are given below.

Parameter	Pre-hepatic/ haemolytic jaundice	Hepatocellular/ hepatic jaundice	Post hepatic/ obstructive jaundice
Conjugated bilirubin	Normal	+/–	++
Unconjugated bilirubin	++	+/–	Normal
ALP	Normal	2–3 times increased	10–15 times increased
AST, ALT	Normal, AST is mildly elevated	>3 times increased	<3 times

Parameter	Pre-hepatic/ haemolytic jaundice	Hepatocellular/ hepatic jaundice	Post hepatic/ obstructive jaundice
GGT	Normal	+/–	+++ in intra hepatic cholestasis
Bile salts in urine	Absent	absent	present
Urobilinogen in urine	+++	+/–	Absent
Stool colour	Dark	–	Clay coloured

The normal **serum total bilirubin concentration is 0.2 to 1.0 mg/dL.** Of this, the **unconjugated bilirubin is about 0.2–0.6 mg/dL** and the **conjugated** is **0.2–0.4 mg/dL**. When bilirubin in the blood exceeds 1 mg/dL, it is considered to be **hyperbilirubinemia**. It occurs due to the production of more bilirubin than the liver can excrete, or due to the failure of a diseased liver to excrete bilirubin produced in normal amounts, or it may be due to a mechanical obstruction in the bile duct that carries bile from the liver to the intestine. In all these conditions, bilirubin accumulates in the blood, and when it exceeds 3 mg/dL, it diffuses into the tissues which then become yellow. The condition is called **jaundice** or **icterus**.

Jaundice is usually manifested as yellowish discolouration of the sclera (white of the eye), conjunctiva, skin and mucous membranes. The sclera, in particular, is examined for discolouration (reflecting even slight increase in serum bilirubin) because it has a high affinity for bilirubin deposition due to its high elastin content. Students must note that **jaundice is a symptom of an underlying disease and is not a disease in itself.**

Interpretation of laboratory data

Given below are some of the findings of biochemical investigations in a patient. Arrive at a probable diagnosis in each case.

Patient 1

Van den Berg reaction – Indirect positive
Total bilirubin – 11.4 mg/dL
Conjugated bilirubin – 0.6 mg/dL
Unconjugated bilirubin – 10.8 mg/dL

Ans: Hemolytic jaundice

Patient 2

Van den Berg reaction – Biphasic
Total bilirubin – 12.8 mg/dL
Conjugated bilirubin – 6.4 mg/dL
Unconjugated bilirubin – 6.4 mg/dL
ALT – 286 IU/L
AST – 264 IU/L

Ans: Hepatocellular jaundice

Patient 3

Van den Berg reaction – Direct positive
Total bilirubin – 15.6 mg/dL
Conjugated bilirubin – 14.2 mg/dL
Unconjugated bilirubin – 1.4 mg/dL
ALT – 38 IU/L
AST – 20 IU/L
ALP – 450 IU/L

Ans: Obstructive jaundice

Note: Jaundice is discussed further under liver disorders below. Refer to liver function tests based on its excretory functions.

10. Liver Disorders

	Biochemical test	Rationale of the test
1.	• Serum: bilirubin – total, conjugated and unconjugated • Urine: bile pigments, bile salts and urobilinogen	Tests of hepatic excretory function
2.	**Enzyme panel** • Alanine aminotransferase (ALT) • Aspartate aminotransferase (AST) • Alkaline phosphatase (ALP) • Gamma glutamyl transferase (GGT) • 5′ Nucleotidase	Test for hepatic cell injury and bilary tract injury
3.	• Total proteins – Serum albumin, globulin, A/G ratio • Prothrombin time • Blood ammonia	Tests for synthesis function
4.	• Serum hyaluronic acid • Procollagen type I carboxy terminal peptide • Procollagen type III amino terminal peptide • Matrix metalloproteinases • Transforming growth factor beta-1	Direct and indirect markers of hepatic fibrosis

Functions of the Liver

The liver performs diverse functions, namely

- **Metabolic functions**: receives, processes and stores materials absorbed from the digestive tract such as amino acids, fatty acids, glucose, cholesterol, vitamins and minerals.
- **Excretory functions**: helps in bile acid synthesis from cholesterol and the excretion of these acids through bile. Bile pigments and bile salts are also excreted in bile into the intestine.
- **Protective functions**: The Kupffer cells of the liver perform phagocytosis on foreign substances.
- **Detoxification of exogenous compounds**: The liver is the main site for detoxification of drugs and toxins. Ammonia is detoxified to urea.

- **Hematological functions**: The liver performs functions such as the formation of blood, particularly in the embryo, plasma protein synthesis and the destruction of erythrocytes. Serum albumin regulates the plasma oncotic pressure, which modulates the fluid between intravascular and tissue spaces.
- **Site for catabolism**: The liver is the major site for the catabolism of thyroid, steroid and other hormones.
- **Site for storage**: Glycogen, vitamin A, vitamin D, vitamin B_{12} and trace element iron are stored in the liver.

Common Disease Processes that Affect the Liver

The most common disease processes that affect the liver are hepatitis, cirrhosis and tumours. Damage to liver cells is usually caused by **viral infections, toxins** and the **long-term abuse of drugs**, e.g., paracetamol, which produce inflammatory changes resulting in **hepatitis. Chronic alcoholism, autoimmune disease** and persistent viral infection are some of the conditions that lead to the death of liver cells and their replacement by fibrous tissue, resulting in reduced functional liver mass. The condition is called **cirrhosis.** Metastases from the lungs and GI tract cancers reach the liver and form secondary **tumours.**

Liver Function Tests

I. **Tests based on liver excretory function**
   (i) Serum-bilirubin: Total, conjugated and unconjugated
   (ii) Urine: Bile pigments, bile salts and urobilinogen
   (iii) Bromsulphthalein test

II. **Tests on liver enzymes**
   (i) Alanine aminotransferase (ALT)
   (ii) Aspartate aminotransferase (AST)
   (iii) Alkaline phosphatase (ALP)
   (iv) Gamma-glutamyl transpeptidase (GGT)
   (v) 5′ nucleotidase

III. **Tests for synthesis function of the liver**
   (i) Total proteins
   (ii) Serum albumin, globulins and A/G ratio
   (iii) Prothrombin time

IV. **Special tests**
   (i) Ceruloplasmin
   (ii) Ferritin
   (iii) α_1-Antitrypsin
   (iv) α-Fetoprotein

Liver function tests based on clinical classification

Markers of liver dysfunction

(i) Serum bilirubin: Total and conjugated
(ii) Urine: Bile pigments, bile salts and urobilinogen

(iii) Total protein, serum albumin and A/G ratio
(iv) Prothrombin time
(v) Blood ammonia

Markers of hepatocellular injury

(i) Alanine aminotransferase (ALT)
(ii) Aspartate aminotransferase (AST)

Markers of cholestasis (obstruction to the flow of bile)

(i) Alkaline phosphatase (ALP)
(ii) Gamma-glutamyl transpeptidase (GGT)

The various tests reflecting the different functions of the liver are described below.

I. Tests based on excretory function

Endogenously produced compounds such as *bilirubin* and *bile acids* and exogenous compounds such as *indocyanine green* or *aminopyrine* are excreted either into bile or urine.

(i) Bilirubin

Bilirubin is an orange-yellow coloured pigment which is the excretory **end product of heme degradation**. It is conjugated in the liver to form bilirubin diglucuronide and excreted in bile or urine.

1. **Serum bilirubin:** Normal serum bilirubin level varies from **0.2 to 1.0 mg/dL.** Of this, the **unconjugated bilirubin (bilirubin–albumin complex)** varies from **0.2 to 0.4 mg/dL** and **conjugated bilirubin** (diglucuronides 75% and monoglucuronides 25%) varies from **0.2 to 0.6 mg/dL**. A rise in serum bilirubin above 1 mg/dL is abnormal and if it rises to 2 mg/dL, it is manifested as jaundice.

   van den Bergh reaction: In this test, diazotised sulphanilic acid (sulphanilic acid in HCl and sodium nitrite) reacts with bilirubin to form a purple-coloured complex, azobilirubin. **Normal serum does not give a positive van den Bergh reaction**. Bilirubin is not soluble in water, but conjugated bilirubin is.

   When bilirubin is **conjugated**, a purple colour is produced immediately on mixing with the reagent. This response is said to be **van den Bergh direct positive**. When bilirubin is **unconjugated,** the colour is obtained only when alcohol is added. The response is then said to be **van den Bergh indirect positive**. If both conjugated and unconjugated bilirubin are present in increased amounts, a purple colour is produced immediately, and the colour is intensified on adding alcohol. Then the reaction is called biphasic.

Clinical significance of serum bilirubin

- In **hemolytic jaundice, unconjugated bilirubin is increased**. Hence, the van den Bergh test is **indirect positive.**
- In **obstructive jaundice, conjugated bilirubin is elevated**. Hence, the van den Bergh reaction is a **direct positive.**
- In **hepatocellular jaundice,** a **biphasic** reaction is observed since both **conjugated and unconjugated bilirubins are increased.**

Jaundice is the most common characteristic clinical manifestation of fluctuations in serum bilirubin. It is caused by an elevated serum bilirubin concentration of **above 2 mg/ dL.** There may be other signs of bilirubin retention such as the passage of tea-coloured urine or tan-coloured stools. *Jaundice is not specific to liver disorders* and may indicate other disorders, including hemolysis and disorders of bilirubin metabolism. Elevated levels of bilirubin can cause **kernicterus** in neonates.

There are three types of jaundice: (1) hemolytic jaundice (pre-hepatic jaundice); (2) obstructive jaundice (post-hepatic jaundice) and (3) hepatic jaundice (intra-hepatic jaundice). The causes for each of these are different. These three types of jaundice are characterised by the elevation of serum bilirubin levels, conjugated or unconjugated or both types. These elevations are assessed by performing different types of tests.

(ii) Urinary bilirubin and urobilinogen

Bilirubin is synthesised from heme in the liver and then transported to the kidney for excretion. Bilirubin is conjugated in the liver, so only this bilirubin is excreted in the urine. Unconjugated bilirubin is not excreted in the urine because of its insolubility. The following tests assess urinary bilirubin and urobilinogen.

(a) **Fouchet's test**: It is used to detect urinary bilirubin. In this test, $BaCl_2$ and $MgSO_4$ are added to urine in a test tube. $BaSO_4$ is precipitated, which then binds to any bilirubin present in the urine. Fouchet's reagent is then added, which oxidises bilirubin to a green or blue colour.
(b) **Ehrlich's test**: It is used to detect urobilinogen in urine.
(c) **Gmelin's test**: Bilirubin in urine can also be detected by this test.

Clinical significance of urinary bilirubin and urobilinogen

In case of jaundice, urine should be examined for the presence of bilirubin, bile salts and urobilinogen. Only conjugated bilirubin is soluble in water and is excreted in the urine. Since the unconjugated bilirubin is increased in blood in prehepatic jaundice, it does not appear in urine.

In obstructive jaundice, bilirubin is conjugated, and is not excreted through the normal passage due to obstruction. Hence, it is regurgitated back into the blood stream, which is then excreted through urine. In cases of obstruction, bile does not reach the intestine and so, urobilinogen may be decreased or absent in urine.

In hepatocellular jaundice, urobilinogen is initially elevated but decreases or disappears later when there is obstruction. It then reappears when the obstruction is cleared. The detection of urinary bilirubin and urobilinogen by simple tests are therefore useful in following the course of a case of jaundice.

Urobilinogen is absent in urine when there is obstruction to bile flow. The first indication of recovery is the reappearance of urobilinogen in urine. It is increased in urine in hemolytic anemias.

(iii) Bile salts

Bile salts are sodium or potassium salts of glycocholates and taurocholates. These are formed in the liver from cholesterol. They reduce the surface tension and help in the digestion and absorption of fats.

Hay's test to assess bile salts: In this test, urine is taken in a test tube and a little **sulphur** powder is sprinkled on it. If the sulphur sinks to the bottom, it indicates that bile salts are present in the urine.

Clinical significance of bile salts: Bile salts are excreted in urine in obstructive jaundice due to bile duct obstruction by inflammation or due to gallstones in the gall bladder.

Bromsulphthalein test

This test is based on the excretory function of the liver. When bromsulphthalein (BSP) is injected, it is mostly excreted through bile and only less than 2% is excreted by the kidneys. Hepatic cells take up bromsulphthalein and conjugate it with glutathione. A significant fraction of BSP is excreted unconjugated. A single bolus dose of BSP, that is, 50 g/L solution is given and the serum concentration is measured at 45 minutes and at two hours' time.

In normal people, the retention of BSP at 45 minutes is less than 5%. The value is slightly higher in old age. The test is very sensitive. **Impairment of liver cell function will cause an increase in BSP retention.** Hence, it is very useful to detect mild impairment. The rate of uptake of BSP by hepatocytes exceeds the excretory capacity, and as a result, it is retained in the liver. Increased plasma retention can result from decreased excretory rate as seen in **Dubin–Johnson syndrome**.

II. Tests based on metabolic function

The liver plays an important role in the regulation of carbohydrate, lipid and protein metabolisms. A bidirectional flux of precursors and products such as glucose, amino acids, free fatty acids and other nutrients take place across the hepatocyte membrane.

Normal blood glucose concentrations are maintained during short fasts by the breakdown of hepatic glycogen and during prolonged fasting by hepatic gluconeogenesis (from amino acids derived from muscle proteins).

In human beings, oxidation of odd-numbered fatty acids also occurs in the liver. But substances that are exclusively metabolised in the liver are very few in number. One among them is **galactose.** Liver function can be assessed by measuring the utilisation of galactose.

(i) Galactose tolerance test

In this test, 350 mg galactose/kg body weight as 25%–30% solution is given intravenously within 3 minutes. Then, blood samples are collected at 10-minute intervals for 1 hour. The half-life of galactose in the blood is about 10–15 minutes in normal individuals. In addition to this test, 14C-labelled galactose can be given and the elimination of 14C-labelled CO_2 may be assessed. Galactose is markedly elevated in **hepatocellular damage** (infective hepatitis and cirrhosis).

(ii) Plasma amino acid test

The amino acid profile is abnormal in hepatic coma. The levels of aromatic amino acids are increased but branched-chain amino acids are decreased.

(iii) Anti-pyrine breath test

Anti-pyrine is rapidly and completely absorbed from the intestine and mostly metabolised by hepatic *Cyt P450 mono-oxygenase system.* 14C-labelled anti-pyrine is given orally and breath samples are collected at 2 and 24 hours. Normal subjects excrete 5%–8% of the administered dose in 2 hours; patients with *cirrhosis* excrete only 2%–3%, and those with *hepatitis* only 2%–4%.

III. Tests based on synthesis function

Many plasma proteins and coagulation proteins are synthesised in the liver. Any decrease in the synthesis of these proteins in pathological conditions that affect the liver leads to a decrease in their plasma concentration. Their proportion in the serum can be determined by **serum protein electrophoresis,** which provides clues for the diagnosis of various conditions. Plasma proteins like albumin, prealbumin, ceruloplasmin, α_1-antitrypsin, haptoglobin, α_2-macroglobin, transferrin and α-fetoprotein are synthesised in the liver. Fluctuations in their concentrations can be used in the diagnosis of various liver diseases. Immunoglobulins are not synthesised in the liver, but estimation of their levels can be used to assess liver function indirectly.

1. **Serum albumin:** Serum albumin is quantitatively the most important protein synthesised by the liver and an estimation of its concentrations reflects the functional status of the liver.

   Since albumin has a long half-life of 20 days, it is not a good indicator of *acute liver diseases.* But in all *chronic diseases of the liver,* the albumin level is *decreased.* A **reversal in A/G ratio** is seen in **cirrhosis of the liver** due to the reduced synthesis of albumin and associated elevated ψ-globulins.

   Normal albumin level in the blood is **3.5–5.0 g/dL** and **globulin level** is **2.5–3.5 g/dL.** The turnover rate of albumin is longer than other plasma proteins and hence, it does not reflect any recent changes in liver function.
2. **Ceruloplasmin:** Ceruloplasmin is a copper-containing protein. Its levels are increased in liver diseases such as chronic active hepatitis, biliary cirrhosis and hemochromatosis. Its levels are decreased in Wilson disease.
3. α_1-**Antitrypsin:** Children with α_1-antitrypsin multiple allelic variants are predisposed to liver disease.
4. **Haptoglobin:** The turnover of haptoglobin is low. Hence, it is used to identify any recent changes in liver functions. Low levels of haptoglobin are seen in severe hepatocellular liver disease and hemolytic states.
5. α_2-**Macroglobin:** Serum levels of α_2-macroglobulins are increased in hepatobiliary disease, chronic active hepatitis and alcohol-induced liver cirrhosis.
6. **Transferrin:** Transferrin is an iron transport protein. Its saturation (which is more than 80%) is used as a screening test for hemochromatosis. Reduced values are found in cirrhosis and malnutrition.
7. α-**Fetoprotein:** It is a normal component of fetal blood but disappears a few weeks after birth. Mild elevations are seen in patients with cirrhosis and acute and chronic hepatitis. It is present in higher concentrations in hepatocellular carcinoma and is used as a **tumour marker**.
8. **Prothrombin time (PT):** Prothrombin is synthesised by the liver and hence, is a useful indicator of liver function. The half-life of prothrombin is only 6 hours. Therefore, prothrombin time indicates the functional status of the liver. PT is prolonged only when the liver loses more than 75% of its reserve capacity. Vitamin K deficiency is also a cause for prolonged prothrombin time. In liver disease, the PT remains prolonged even after the administration of vitamin K.
9. **Electrophoretic separation of plasma proteins:** Plasma protein separation through electrophoresis gives valuable information about underlying liver diseases. **Albumin** and **prealbumin** bands are reduced in cirrhosis and acute hepatitis, respectively. α_1 **fraction of globulins** is decreased in hepatocellular disease and increased in malignancies and febrile conditions. α_2-**globulins and** β-**globulins** are increased in biliary obstruction. γ-**globulins** are increased in cirrhosis.

IV. Tests based on enzyme estimations

Liver cells contain many enzymes which are released into the blood when there is hepatic damage. Measuring the levels of these enzymes can be used in the assessment of liver function and in the diagnosis of various liver diseases. These are briefly discussed below.

1. **Aminotransferases or transaminases:** The two aminotransferases, namely **alanine aminotransferase** (**ALT,** previously called serum glutamate pyruvate transaminase- **SGPT**) and **aspartate aminotransferase** [**AST,** previously called serum glutamate oxaloacetate transaminase (**SGOT**)] are widely used to assess liver function. Of these two, serum ALT levels are considered to be more sensitive and specific.

   The levels of aminotransferases in serum are **elevated** in all **liver diseases.** The **normal** concentrations of **ALT** and **AST** in plasma are **10–35 IU/L and 8–20 IU/L respectively**. The degree of elevation in their plasma levels reflects the extent of hepatocellular damage. **Very high levels** (more than 1,000 IU/L) are observed in **acute hepatitis. Moderate elevation** of aminotransferases (100–300 IU/L) is seen in **alcoholic hepatitis, non-alcoholic chronic hepatitis,** autoimmune hepatitis and Wilson disease. **Minor elevation** (less than 100 IU/L) is observed in **fatty liver, chronic viral hepatitis** (hepatitis C) and in non-alcoholic steatohepatitis (NASH). However, in cirrhosis and chronic hepatitis, serum ALT levels do not correlate with the degree of liver cell damage.

   Obese individuals, even though normal, may exhibit increased ALT levels.

   AST/ALT ratio: The normal AST:ALT ratio is 0.8. It is **increased** (>2) in **myocardial infarction, alcoholic hepatitis** and **cirrhosis**. It is **decreased** in acute **hepatocellular injury** and **cholestasis.**
2. **Alkaline phosphatase (ALP):** Alkaline phosphatase (ALP) is produced by bones and the liver (epithelial cells lining the biliary canaliculi and placenta). Bile duct obstruction induces the synthesis of ALP by the biliary tract epithelial cells. Hence, **high levels of ALP** (normal 3–13 KA units/dL) indicate **obstructive/posthepatic jaundice.** ALP is also elevated in **hepatic carcinoma** and **cirrhosis**. Obstructive jaundice (cholestasis) may be extrahepatic or intrahepatic. ALP levels are elevated as high as a dozen times of its normal concentration in extrahepatic cholestasis. It is usually caused by gallstones or by the pressure exerted by carcinoma of the head of the pancreas in the bile duct. Intrahepatic obstruction may be due to viral infection (infective hepatitis) or caused by drugs (e.g., chlorpromazine).

   However, it should be noted that the liver is not the only source of ALP. Hence, the interpretation of ALP levels must be considered along with other liver function tests (LFTs). The liver and bone isoenzymes can be separated by electrophoresis.
3. **Gamma-glutamyl transpeptidase (GGT):** Elevated levels of GGT are associated with **chronic alcoholism.** Its levels are also increased (normal 5–50 IU/L) in **biliary obstruction,** pancreatic disease and chronic obstructive pulmonary disease (COPD). Its activity is usually parallel to that of ALP.
4. **Nucleotidase:** The activity of 5′-nucleotidase (normal 2–15 IU/L) is **elevated** in obstructive liver disease (**hepatobiliary disease**). Its activity is parallel to that of ALP. But unlike ALP, which is also altered in bone disease, nucleotidase is **specific to liver function**.
5. **Other enzymes:** LDH2, an isoenzyme of lactate dehydrogenase, isocitrate dehydrogenase and glutathione-s-transferase are also liver enzymes of diagnostic importance.
6. **Liver enzyme panels:** It should be noted that on most occasions, a combination of serum enzyme estimations is more useful than single estimation in assessing the liver function and arriving at a proper diagnosis. Marked increase in aminotransferases (ALT and AST) relative to a small increase in ALP indicates underlying hepatocellular damage. On the other

hand, a small increase in aminotransferases and a marked increase in ALP indicates biliary obstruction. In other words, **ALT and AST are used as markers of hepatocellular injury, whereas ALP and GGT are used as markers of cholestasis.**

V. Tests based on detoxification function

Blood ammonia test

Blood ammonia estimation serves as an index of urea synthesis by the liver. Raised ammonia in the plasma is suggestive of cirrhosis. Arterial blood should be used for blood ammonia estimation. Estimation of ammonia is used to exclude or diagnose hepatic failure in patients with unexplained stupor or coma. In neonates with suspected urea cycle disorders, ammonia estimation is indicated.

VI. Selection of liver function tests

Liver function tests (LFTs) are the commonly employed biochemical tests carried out to diagnose and monitor the course of a liver disease. Different tests are done for the overall assessment of the liver function.

LFT profile

The most **commonly employed liver function tests** in clinical practice are serum bilirubin, total protein, A/G ratio, albumin, ALP, ALT, AST and GGT. ALT and ALP distinguish the pattern of liver disease (hepatocellular damage or biliary obstruction). Albumin levels indicate whether hepatitis/ cholestasis is acute or chronic (albumin levels are low in chronic conditions). Prothrombin time reflects the degree of the severity of liver dysfunction.

11. Pancreatitis

Serum amylase and serum lipase: Both amylase and lipase enzymes will be elevated due to inflammation and damage to pancreas. Lipase is more specific than amylase and remains elevated for a longer period compared to amylase.

(i) Measurement of Pancreatic Enzymes

Amylase is the enzyme that digests starch. Serum amylase contains P (pancreatic) and S (salivary) isoenzymes. These two can be distinguished by the inhibition test. A protein inhibitor present in the alcoholic extracts of wheat will exclusively inhibit the S isoenzyme. Normal amylase level in serum is 50–120 units. The level rises within five hours of the onset of **acute pancreatitis** and reaches a peak within 12 hours. Within two to four days of the attack, the level returns to normal. As the serum amylase level starts falling, urinary amylase level rises.

Serum lipase is the enzyme that hydrolyses glycerol esters of fatty acids. Its level in blood is highly elevated in **acute pancreatitis** and this persists for 7–10 days, which is longer than that of amylase. Further, lipase is not increased in salivary diseases. Hence, lipase estimation has an advantage over that of amylase.

(ii) Secretin–Cholecystokinin (CKK) Test

The duodenal contents are aspirated in the fasting state. Secretin (1 unit/kg body wt.) is administered, followed by CCK. The duodenal contents are aspirated again at 10-minute intervals for 1½ hours. Each sample is analysed for volume, bicarbonate content and amylase activity. If bicarbonate secretion is more than 15 mmol/L at 30 minutes, the secretory function is considered normal.

12. Acid–Base Disorders

Acidosis is the clinical state where acids accumulate or bases are lost. A loss of acid or accumulation of base leads to **alkalosis.**

Classification of Acid–Base Disturbances

Acidosis (fall in pH)

Acidosis refers to an elevation in arterial plasma H^+ ion concentration and decline in plasma pH.

- **Respiratory acidosis**: Primary excess of carbonic acid
- **Metabolic acidosis**: Primary deficit of bicarbonate

Alkalosis (rise in pH)

Alkalosis refers to a decrease in arterial plasma H^+ ion concentration and a rise in plasma pH.

- **Respiratory alkalosis**: Primary deficit of carbonic acid
- **Metabolic alkalosis**: Primary excess of bicarbonate

Acid–base parameters are to be checked in patients with

- Any serious illness
- Multiorgan failure
- Respiratory failure
- Cardiac failure
- Uncontrolled diabetes mellitus
- Poisoning (barbiturates, ethylene glycol)

Laboratory investigations in acid–base disorders

	Parameters	Rationale
1.	**Arterial blood gas (ABG) analysis**	
	pH	Low pH seen in metabolic acidosis, respiratory acidosis
	pCO_2	High in metabolic alkalosis, respiratory alkalosis
	pO_2	– High in respiratory acidosis – Low in respiratory alkalosis
	HCO_3^-	– Low in respiratory alkalosis. – Low in in metabolic acidosis – High in metabolic alkalosis
	Serum electrolytes	Electrolyte imbalance is seen in ABG disorders, and it also helps in calculating anion gap.

	Parameters	Rationale
2.	– **Random blood glucose** – **Renal function tests** – **Blood and urine test for identifying any metabolic disorders**	To assess the underlying cause

Clinical case studies

1. A 60-year-old man was brought to hospital in a semiconscious state. The patient displayed a typical hyperventilatory breathing pattern with a sweetish odour of breath. His pulse was feeble and BP was low. Blood gas analysis and other laboratory investigations were performed and results obtained as follows.

Parameter	Obtained value	Reference range
Blood pH	7.1	7.35–7.45
Plasma HCO_3^-	11 mmol/L	20–30 mmol/L
pCO_2	38 mmHg	35–45 mm Hg
Blood glucose (random)	550 mg/dL	less than 140 mg/dL

Benedict and Roth's tests were positive in the urine samples. How would you interpret this case and what is your probable diagnosis?

Probable diagnosis: It is a case of **severe diabetes mellitus** left untreated resulting in *diabetic ketoacidosis*. It is a classic example of **metabolic acidosis**. The blood glucose levels are very high and this is associated with urinary excretion of glucose and ketone bodies. The patient has typical rhythmic respiration (*Kussmaul's respiration*). The fruity odour of the breath is due to the presence of acetone in the exhaled air. Note that pCO_2 values are within normal limits as it is compensated by hyperventilation. The plasma concentration of HCO_3^- is markedly low. This is the typical finding of non-respiratory acidosis. Bicarbonate levels are depleted as they have been consumed to buffer H^+ ions. The reduced HCO_3^- concentration is the primary abnormality in metabolic acidosis and the plasma HCO_3^- levels reflect the degree of severity of the metabolic acidosis.

2. A 50-year-old man was admitted to hospital with a complaint of persistent vomiting. On examination he was found dehydrated and his respiration was shallow. He gave a past history of dyspepsia. The results of the laboratory investigations are as follows.

Parameter	Obtained value	Reference range
Blood pH	7.72	7.35–7.45
Plasma HCO_3^-	45 mmol/L	20–30 mmol/L
pCO_2	60 mmHg	35–45 mm Hg
Na^+	140 mEq/L	136–145 mEq/L
K^+	2.5 mEq/L	3.5–5 mEq/L

How can the interpretation of the above laboratory data aid you in your diagnosis?

Ans: Elevated blood pH and an increase in plasma HCO^- levels are characteristic of **metabolic alkalosis**. The increase in pCO_2 is a result of compensatory hypoventilation leading to retention of carbon dioxide. Dehydration by fluid loss stimulates Na^+ reabsorption in the kidneys. The patient excretes more H^+ and K^+ ions in exchange for the reabsorption of Na^+. This is why the patient excretes acidic urine despite alkalosis.

3. A young man who had met with a road accident that caused injury to his chest wall was brought to the casualty ward of a hospital. The results of the laboratory investigations carried out on him are as follows.

Parameter	Obtained value	Reference range
Blood pH	7.2	7.35–7.45
Plasma HCO_3^-	26 mmol/L	20–30 mmol/L
pCO_2 60 mmHg	35–45 mm Hg	

Interpret the laboratory data and diagnose.

Ans: The low pH and high pCO_2 values are suggestive of **respiratory acidosis**. The almost normal HCO^- level indicates that no renal compensation has occurred as yet. Renal compensation needs several days to become fully effective, in contrast to the respiratory compensation which occurs within few minutes in metabolic acid–base disorders.

4. A medical student volunteered a sample of arterial blood for a hematology experiment in the physiology lab. The tutor took some time to explain the procedure to the class, during which time, the student became increasingly anxious and started complaining of a tingling feeling and numbness in his fingers and toes. The results of the investigations of the arterial blood are as follows.

Parameter	Obtained value	Reference range
Blood pH	7.5	7.35–7.45
Plasma HCO_3^-	24 mmol/L	20–30 mmol/L
pCO_2	25 mmHg	35–45 mm Hg

What is your assessment of the case?

Ans: There is an increase in blood pH and lowered pCO_2 indicating that the student was becoming alkalotic. Going by the reduced pCO_2, the patient's condition can be labelled as a **respiratory alkalosis**. The low pCO_2 is a result of anxiety-induced hyperventilation and no compensation has occurred in such a short time span. The symptoms of tingling and numbness are associated with a decline in the calcium ion levels, which is common in alkalosis.

13. Thyroid Disorders

Tests	Biochemical basis of the tests
T_3, T_4 TSH, FT_3, FT_4	1. **Primary hypothyroidism** – High TSH, low FT_3 and FT_4 *Causes:* Lack of iodine (intake), autoimmune thyroid destruction (Hashimoto thyroiditis), lack of feedback inhibition from thyroid hormones (TSH increases) 2. **Primary hyperthyroidism** – Low TSH, High FT_3 and FT_4 *Causes*: Graves disease, initial stages of Hashimoto thyroiditis causes transient increase in FT_3 and FT_4
Anti-TPO (thyroperoxidase) antibodies	Hashimoto thyroiditis
Thyroid receptor antibodies, thyroglobulin antibody	Graves disease
Lipid profile	Hypercholesterolemia is seen in hypothyroidism. Hypocholesterolemia is seen in hyperthyroidism.
Increased glucose in plasma	Increased / decreased metabolism in hyperthyroidism, hypothyroidism leads to changing plasma glucose levels.

Disorders of Thyroid Function

The most common disorders of thyroid function are hyperthyroidism (thyrotoxicosis), hypothyroidism and goitre.

Goitre

Any enlargement of the thyroid gland is called goitre. The size of the thyroid gland increases to compensate for the decreased synthesis of thyroid hormones and is always associated with elevated levels of TSH. Goitre is primarily due to a failure in the autoregulation of T_3 and T_4 caused by either the deficiency or excess of iodide. Rare inherited metabolic defects anywhere along the biosynthetic pathway of thyroid hormones may also cause goitre.

In geographical areas where the iodine content of the soil is very low, simple endemic goitre may be seen. Iodised salt intake is advised in such cases. Drugs known to interfere with thyroid hormone synthesis such as thiourea, perchlorates and thiocyanates may cause goitre.

Hyperthyroidism

Hyperthyroidism (overactivity of the thyroid gland) is seen in about 0.5% of the general population. It can be caused by a number of conditions. The most common cause of hyperthyroidism is **Graves disease,** which is an autoimmune disease occurring due to the presence of antibodies which bind and stimulate the TSH receptors in the thyroid gland. These autoantibodies, known as **thyroid-stimulating IgG** or **long-acting thyroid stimulators (LATS),** cause excess production of T_3 and T_4 by the activation of TSH.

The signs and symptoms of hyperthyroidism include elevated BMR, tachycardia, menstrual irregularities, sleeplessness, weight loss in spite of increased appetite, heat intolerance, palpitations, restlessness, excessive sweating, tremors and muscle weakness. Biochemical diagnosis depends on suppressed TSH and raised free T_3 and T_4 levels. The management includes treatment with anti-thyroid drugs, and in severe forms the surgical removal of the thyroid gland is indicated.

Hypothyroidism

Compared to hyperthyroidism, hypothyroidism is more common and occurs in 2–15 percent of the general population in India. It is more common in women than in men and the incidence increases with age. The clinical features of hypothyroidism include tiredness and lethargy, intolerance of cold conditions, weight gain, hoarseness of voice, hair loss, constipation, depression, bradycardia, dry and coarse skin, elevated cholesterol levels and carpal tunnel syndrome.

Since thyroid hormones are required for normal development, neonatal hypothyroidism may result in **cretinism** manifested as multiple congenital defects and mental retardation. Hypothyroidism in adults causes **myxedema** with symptoms such as puffiness of the face, bagginess under eyes, slowness in physical activities and overall weight gain.

The biochemical diagnosis of hypothyroidism will reveal high TSH levels and low free T_4. Elevated TSH levels in the presence of normal T_4 is known as **subclinical hypothyroidism,** a condition that may lead to overt hypothyroidism at a later time.

Thyroid Function Tests

For long, the measurement of BMR and the estimation of PBI (protein-bound iodine) were used to assess the thyroid function. Nowadays, these have been completely replaced by the direct measurement of thyroid hormones by advanced techniques such as RIA or ELISA. Measurement of antibodies against the TSH receptor or other thyroid antigens are also done. The normal reference values of serum levels of thyroid hormones are given in Table 27.3.

Table 27.3 *Reference values of thyroid hormones*

Estimated hormone	Reference value
Total T_3	1.2–2.1 nmol/L or 77–135 ng/dL
Total T_4	70–150 nmol/L or 5.4–11.7 µg/dL
Free T_3	3.7–6.5 pmol/L or 2.4–4.2 pg/dL
Free T_4	9.0–16 pmol/L or 0.7–1.24 ng/dL
Reverse T_3	0.15–0.4 nmol/L or 10–25 ng/dL
TSH	0.34–4.25 µIU/L

(i) Estimation of T_4 and T_3

Estimating the total T_4 (thyroxin) and T_3 (triiodothyronine) by radioimmunoassay (RIA) or enzyme linked immunosorbent assay (ELISA) techniques provide diagnostic clues about thyroid function. Both T_4 and T_3 are decreased in hypothyroidism and increased in hyperthyroidism. Commonly, the changes in serum T_3 are proportionally higher than the changes in T_4.

When the results of total T_4 and T_3 are borderline, it is advisable to estimate the levels of free T_4 and T_3. Students may recall that thyroid hormones in blood occur in two forms: (i) protein-bound, and (ii) free form. Since thyroid hormones induce biological actions in the free form, any increase or decrease in their levels provides a more accurate diagnosis of thyroid disease. Reference values of thyroid hormones are as given in Table 27.3.

(ii) Estimation of TSH

Thyroid disorders are of two types:

- Primary (defect in thyroid gland)
- Secondary (defect occurs in hypothalamus or pituitary gland). Estimation of serum TSH levels are useful in differentiating between these two types.

In hypothyroidism of thyroid origin **(primary hypothyroidism)** the TSH level is elevated because of the lack of feedback effect, and in **primary hyperthyroidism** serum TSH is decreased due to negative feedback by T_3 and T_4.

Likewise, in **secondary hypothyroidism** (of pituitary or hypothalamic origin) the TSH levels are low, whereas in **secondary hyperthyroidism** high TSH levels are observed.

(iii) Radioactive iodide uptake

The ability of the thyroid gland to trap iodide is studied by administering a very small dose of radioactive iodine (5 µg of ^{131}I) orally. Radioactivity is measured at different time intervals (usually at 6 hours and again at 24 hours). In healthy persons, 10–30% of the ingested iodide is detected within 24 hours. **The uptake is increased in hyperthyroidism and decreased in hypothyroidism.**

(iv) Thyroid-hormone-binding proteins

Most of the circulating T_3 and T_4 are bound to certain binding proteins such as **thyroxine-binding globulins (TBG)**. Hence, any alterations in the levels of these binding proteins affect the plasma concentrations of T_4 and T_3. The amount of thyroid hormones transported will increase with increased TBG, or decrease with decreased TBG. Likewise, decreased hormone-binding affinity liberates thyroid hormones from TBG, resulting in decreased total serum T_4 and T_3, and increased free T_4 and free T_3.

(v) TRH stimulation test or TRH response test

In normal healthy individuals, the administration of TRH causes increased secretion of TSH, and consequently, of thyroid hormones.

In **hypothyroidism**, TRH stimulation causes a release of high levels of TSH. This is due to the reduced feedback inhibition by the low levels of thyroid hormones. In **hyperthyroidism**, feedback inhibition by the high levels of thyroid hormones surpasses the stimulating effect of TRH on TSH, resulting in a blunted TSH response. Blunted TSH response is also seen when the receptors for TRH on pituitary are defective.

(vi) Other thyroid function tests

Detection of **autoantibodies** [(**thyroid-stimulating IgS** or **long-acting thyroid stimulators (LATS)**] to the thyroid gland indicate excess production of T_3 and T_4 by the activation of TSH, resulting in hyperthyroidism.

Basal metabolic rate (BMR) is increased in hyperthyroidism and decreased in hypothyroidism. **Serum cholesterol** levels are decreased in hyperthyroidism and are increased in hypothyroidism. These two parameters, even though not directly related to the thyroid gland, provide a clue about thyroid status.

Clinical case study

A 32-year-old woman presented with infertility after 7 years of marriage. She gives a history of easy fatigue, weight gain, muscle pains, constipation and cold intolerance. On examination she is found to be anemic. Laboratory results are as follows:

Investigation	**Result**	**Reference range**
Free T3	80 ng/dL	80–220 ng/dL
Total T4	5 mg/dL	5–12 mg/dL
TSH	20 µIU/L	<5 µIU/L
Cholesterol	300 mg/dL	<200 mg/dL

Probable diagnosis: The clinical signs and symptoms of the woman with elevated levels of TSH and relatively low T_4 and T_3 are suggestive of **hypothyroidism**. Cholesterol levels are usually elevated in patients of hypothyroidism.

14. Vitamins and Minerals

The biochemical tests for ascertaining vitamin and mineral deficiencies manifesting as a variety of disorders are outlined below.

Vitamin / Mineral	Tests	Rationale
Vitamin A	Serum retinol binding protein	Decreases in vitamin A deficiency
	Serum vitamin A	Decreases in vitamin A deficiency
	Serum LFT	Secondary vitamin A deficiency associated with liver disorder
Vitamin D	Serum vitamin D_3	Decreases
	Calcium	Slightly low
	Serum phosphate	Serum ALP markedly increases (in osteomalacia)
	RBS in pregnant women	Gestational diabetes mellitus is associated with low levels of vitamin D
	Serum LFT	Vitamin D deficiency associated with liver disorder
Vitamin K	Prothrombin time (PT) and clotting time (CT)	Prolonged PT and delayed CT are characteristics of vitamin K deficiency.
	Serum LFT	Defective synthesis of coagulation factors. Obstructive jaundice can also cause vitamin D deficiency.

Vitamin / Mineral	Tests	Rationale
Thiamine	Serum LFT	Thiamine deficiency is seen in chronic alcoholics.
	Blood thiamine, blood pyruvate	Reduced
	α-Ketoglutarate and lactate	Increases in thiamine deficiency
	Erythrocyte transketolase	Erythrocyte transketolase activity is reduced (Earliest manifestation)
Riboflavin	Glutathione reductase activity in RBC	Riboflavin deficiency can be measured by glutathione reductase activity as it is FAD dependent
	LFT in newborns	In neonatal jaundice, riboflavin will be destroyed during phototherapy
	Tests for thiamine and niacin deficiency	Riboflavin deficiency will be accompanied by other deficiencies like beri beri and pellagra
	Urinary riboflavin	Increased
Niacin	Urinary niacin	Increased
	5-Hydroxyindole acetic acid in urine (5-HIAA)	Niacin deficiency is seen in carcinoid syndrome.
	Urinary N-methyl nicotinamide and Pyridine.	Decreased in niacin deficiency
Pyridoxine (B_6)	Plasma B_6	Reduced
	Plasma PLP (Pyridoxal 5' phosphate)	Decreased
	Serum LFT	Deficiency seen in chronic alcoholics. AST and ALT decreased
	Measurement of urinary tryptophan metabolites (Xantho urinary acid)	In B_6 deficiency, xanthourinary acids excretion is increased.
Folic acid	Serum folate	Decreased deficiency. Macrocytic anemia is seen in folate deficiency.
	Plasma homocysteine	Increased in folic acid deficiency
	FIGLU excretion test	In folate deficiency, FIGLU is excreted in urine.

Vitamin / Mineral	Tests	Rationale
Folic acid	AICAR excretion test	Amino imidazole carboxamide ribose 5 phosphate (AICAR) accumulates and is excreted in urine in folate deficiency.
	Peripheral smear	Macrocytic anemia is seen in folate deficiency.
Cyanocobalamin	Serum B_{12}	Reduced
	Schilling test	To test B_{12} absorption
	Urinary methyl malonic acid test	Methyl malonic acid Increases in urine in B_{12} deficiency
	Plasma homocysteine	Increased in B_{12} deficiency
	Peripheral smear	Deficiency shows megaloblastic anemia
Vitamin C	Serum ascorbic acid	Low in vitamin C deficiency
Calcium	Serum calcium	Increased in hypercalcemia and decreased in hypocalcemia
	Serum vitamin D_3	Deficiency of vitamin D causes hypocalcemia.
	Parathyroid function tests	Associated with calcium homeostasis
	Serum RFT	Defective formation of calcitriol in renal diseases.
	Serum magnesium	Hypomagnesimea is associated with hypocalcemia
	Thyroid function tests	Thyrotoxicosis causes hypercalcemia
	LFT	ALP may be increased in hypercalcemia
Phosphate	Serum phosphate (early morning fasting sample)	Decreased in hyperparathyroidism and rickets
	Thyroid function tests	In thyrotoxicosis, increased serum phosphate
	RFT	Hyperphosphatemia seen in renal impairment.
	Parathyroid function tests	In hyperparathyroidism, increased urinary excretion of phosphate
	Serum vitamin D_3	Vitamin D deficiency causes hypophosphatemia
	Serum calcium	Hypercalcemia causes hypophospatemia

Vitamin / Mineral	Tests	Rationale
Iron	Serum ferritin	In iron deficiency anemia, ferritin content is reduced.
	Serum iron level	Depressed in iron deficiency
	Total iron binding capacity (TIBC)	Increased
	Thyroid function tests	Iron level decreased in hypothyroidism
Copper	Serum/ plasma copper	Low in deficiency
	Serum ceruloplasmin	This copper-containing protein decreases in copper deficiency
	LFT	Liver is an important organ for storing copper. Abnormal copper metabolism is seen in individuals with liver disease.
Magnesium	Serum magnesium	Low in magnesium deficiency
	Serum potassium and calcium	Unexplained hypokalemia or hypocalcemia suggests magnesium deficiency
	Thyroid function tests	Hypothyroidism can cause low magnesium
	LFT	Chronic alcoholism can cause low magnesium level.
Iodine	Urinary iodine	Iodine taken up is either converted to thyroid hormone or excreted in the urine.
	Iodine induced thyroid disease (Thyroid function tests)	
Zinc	Plasma zinc, zinc concentration in neutrophils	Decreased
	Serum albumin	Lowered plasma zinc concentration is seen in hypoalbuminemia
	LFT	In hepatic cirrhosis zinc deficiency is seen. ALP is decreased.

Viva Questions

1. Describe the various lab investigations that play a role in the diagnosis of diabetes mellitus.
2. What is the significance of glycated hemoglobin?
3. Describe briefly the laboratory analytes of lipid metabolism.
4. What is Friedewald equation?

5. What are the laboratory investigations that help in diagnosing myocardial infarction?
6. What is glomerulonephritis?
7. How is the laboratory diagnosis done for gouty arthritis?
8. What is the diagnostic significance of proteinuria?
9. How do you interpret laboratory data in the diagnosis of different types of jaundice?
10. Name the liver function tests (LFT) based on various functions of liver.
11. What is secretin-cholecystokinin test?
12. Name the different types of acid-base disorders. What are the parameters that help in their diagnosis?
13. List the thyroid function tests and explain briefly their role in diagnosis of thyroid diseases.
14. What are the enzyme markers of pancreatitis?
15. What is the biochemical mechanism underlying edema?
16. What are the biochemical diagnostic markers in myocardial infarction?
17. Enumerate the diagnostic tests related to various vitamins.
18. Name the diagnostic tests related to various minerals.

28 Urinary Screening for Inborn Errors of Metabolism

BI 11.5

Inborn errors of metabolism are due to deficiency or absence of an active enzyme in a metabolic pathway resulting in:

- Decreased synthesis of an important metabolite
- Accumulation of substrate
- Conversion of substrate into other, often undesirable compounds

A newborn baby with an inborn error of metabolism may present with seizures, recurrent vomiting, abnormal smell of urine, atypical facial features, growth retardation, delayed milestones, enlargement of organs, brain damage, or sometimes death. In most countries of the world, newborns are screened at birth for a variety of disorders. These include a group of tests referred to as metabolic screen which includes

- Complete blood picture, blood grouping
- Blood glucose, lactate, ammonia, and uric acid levels
- Liver function tests, renal function tests
- Urine for reducing sugars, ketone bodies.

Based on the clinical presentation or relevant clinical history of the parents, additional screening tests are performed on blood and urine. The blood sample collected from a heel prick is dried as a spot on a filter paper. Urine sample is also collected for analysis. Several screening tests are available to detect a suspected case, but the final diagnosis is only possible by measuring the enzyme levels/activity in the cultured cells from the newborn (fibroblast) or cultured cells from the amniotic fluid (prenatal diagnosis). Molecular genetic techniques and tandem mass spectrometry (TMS) are increasingly being adopted to diagnose inheritable metabolic diseases. A few of the inborn errors of metabolism and their diagnosis are enumerated in the table below.

Glucose-6-Phosphate Dehydrogenase (G6PD) Deficiency

G6PD deficiency is an X-linked recessive disorder. G6PD is an enzyme of the hexose-monophosphate shunt.

Experiment	Observation	Inference
Bernstein method Add sodium nitrite and methylene blue to a citrated blood sample.	The time required for the dye to change to reduced form (colourless) is noted, which may be more than 2 hrs in G6PD deficient individuals.	G6PD deficiency

Principle of the test: Sodium nitrite changes Hb (Fe^{2+}) to Hb (Fe^{3+}). If there is adequate NADPH in the cells, Hb (Fe^{3+}) gets converted back to Hb (Fe^{2+}) using methylene blue. Deficiency leads to a persistence of the brown colour.

Pentosuria

This occurs due to xylitol dehydrogenase deficiency, which leads to high levels of L-xylulose in urine. One of the test for pentosuria is given below. Other tests include genetic testing for confirmation of diagnosis.

Experiment	Observation	Inference
Bial test: Add Bial's reagent (orcinol, HCL, ferric chloride) to the urine sample.	A blue green colour indicates the presence of pentose	Pentose is present.

Galactosemia

Galactosemia manifests as the following types.

Type of galactosemia	Deficient enzyme	Clinical manifestations
Type I (classical galactosemia)	Deficiency of galactose-1-phosphate uridyl transferase	Cataract, mental retardation and cirrhosis
Type II	Deficiency of galactokinase	Principally presents as cataract
Type III	UDP-galactose 4- epimerase deficiency	Cataract

Tests for Galactose in Urine

Experiment	Observation	Inference
Benedict's test	A reddish-brown colour is obtained	Reducing sugars may be present.
Mucic acid test	Crystals of mucic acid is seen	Galactose or lactose is present.
Methyl amine test	Red colour Yellow colour	Galactose is present
Barfoed test	Brick red scum	Monosaccharide may be present. Disaccharide is absent.

Galactose can be identified in urine by chromatography. Also, enzyme deficiency can be estimated in RBCs.

Inherited Disorders of Mucopolysaccharides

Mucopolysaccharidosis is a broad spectrum of disorders due to deficiency of a group of enzymes which degrade three classes of mucopolysaccharides: the dermatan, keratin and heparan sulphate.

Experiment	Observation	Inference
Alcan blue spot test in urine	Blue spot appears	Mucopolysaccharides are present (conc.>100 μg/dL)

Another screening test for mucopolysaccharide is **cetyltrimethylammonium bromide (CTAB) test.**

Phenylketonuria (PKU)

- It is a disorder of phenylalanine metabolism caused due to the deficiency of phenylalanine hydroxylase.
- This enzyme catalyses the conversion of phenylalanine to tyrosine.
- Primary (mutations in phenylalanine hydroxylase gene) or secondary (due to deficiency of co-factor) impairment of phenylalanine hydroxylase results in accumulation of phenylalanine in blood.
- Normal level of blood phenylalanine in 1 mg/dL. In PKU, it is <20 mg/dL.

Phenylketonuria	Defect	Tests
Type I (Classical phenylketonuria)	Phenyl alanine hydroxylase deficiency	– Guthrie' test – Ferric chloride test – TMS (tandem mass spectrometry)
Type I and III (Non-classical phenylketonuria)	Hyper phenylalaninaemia due to tetrahydrobiopterin defect	– Guthrie' test – Ferric chloride test – TMS
Type IV and V (Non-classical phenylketonuria)	Due to defect in enzymes synthesise tetrahydrobiopterin (BH4)	Measurement of neopterin and biopterin enzyme essay tetrahydrobiopterin loading test.

Tests for Phenylketonuria

Experiment	Observation	Inference
Ferric chloride test	Dark green colour develops	Phenylpyruvate is present
Dinitrophenyl hydrazine test (DNPH test)	Appearance of a yellow precipitate	Presence of keto acids

Alkaptonuria

It is an autosomal recessive disorder of tyrosine metabolism due to the deficiency of the enzyme **homogenisate oxidase**. It catalyses the conversion of homogentisic acid to maleyl acetoacetic acid in the tyrosine metabolic pathway.

(*continued*)

Tests for Alkaptonuria in Urine

Experiment	Observation	Inference
Ferric chloride test	Dark green colour develops	Homogentisic acid is present.
Benedict's test	A brown colour is obtained.	Homogentisic acid may be present.

Maple Syrup Urine Disease (MSUD)

This is an autosomal recessive disorder. It is due to deficiency of the branches chain α-keto acid dehydrogenase enzyme leads to a block in the metabolism of branched chain amino acids (valine, leucine, and isoleucine).

Types of MSUD	Enzyme deficiency
Type IA MSUD	Branched chain α-keto acid decarboxylase (α- sub unit)
Type IB MSUD	Branched chain α-keto acid decarboxylase (β- sub unit)
Type II MSUD	Dihydrolipoyltransacylase
Type III MSUD	Dihydrolipoamide dehydrogenase (DLD)

Tests for MSUD

Experiment	Observation	Inference
Ferric chloride test	Greenish grey colour develops, which persists for 2–4minutes.	MSUD may be present.
Dinitrophenyl hydrazine test (DNPH test)	Appearance of a yellow precipitate	Implies the presence of ketoacids

Homocystinuria

It is disorder of sulphur-containing amino acid metabolism with an incidence of 1 in 2,00,000 births. Normal homocysteine level in blood is 5–15 µmol/L. In homocystnuria, it may be increased 50–100 times.

Tests for Homocystinuria

Experiment	Observation	Inference
Cyanide-nitroprusside test	Deep red colour is developed	Cystine and homocystine are present.
To differentiate between cystine and homocystine	1. Immediate development of pink-purple colour 2. Delayed development of pink-purple colour	1. Homocystine is present. 2. Cystine is present.

29

BI 11.23

Energy Content of Various Foods and the Significance of Glycemic Index

To formulate a diet appropriate for different individuals, it is important to calculate the energy content of food items.

Energy Requirements in Humans

	Energy requirement	Protein requirement
Adults male – 70 kg		
Sedentary	2320 kcal	0.83 g/kg
Moderate	2730 kcal	
Heavy	3490 kcal	
Adults female – 70 kg		
Sedentary	1900 kcal	0.83 g/kg
Moderate	2230 kcal	
Heavy	2850 kcal	
Children		
Infants	100 kcal/kg	1.5–2 g/kg
1–5 years	81 kcal/kg	0.94 g/kg
6–10 years	71 kcal/kg	0.91 g/kg
11–18 years (male)	60 kcal/kg	0.88 g/kg
11–18 years (female	55 kcal/kg	0.86 g/kg
Pregnancy additional	350 kcal +	18 g
Lactation additional		
First 6 months	600 kcal	13 g
Next 6 months	520 kcal	8 g

Calorie requirements of a person varies according to

- Basal metabolic rate
- Supply of calories to meet the extra requirement caused by specific dynamic action (SDA)
- Supply of physical activity over and above the basal requirements
- During periods of growth and/convalescence, extra provision has to be made to meet the needs for synthesis of tissues and weight gain.

Energy Content of Foods

The energy content of food is calculated from the heat released during the total combustion of the food in a calorimeter. It is expressed in kilocalories (kcal/g) or joules (J).

The following are the calorie content of various dietary nutrients:

a. Carbohydrates – 4 kcal/g
b. Protein – 4 kcal/g
c. Fats – 9 kcal/g
d. Ethanol – 7 cal/g

List of Common Foods and their Calorie Content

Food	Quantity	Portion/ size	Energy (kcal)	Protein (g)	Fats (g)	Carbohydrates (g)
Boiled rice	150 g	1 Katori	177	3.6	0.3	40
Lime rice	150 g	1 Katori	261	4.2	8.4	42
Curd rice	150 g	1 Katori	158	4.5	3.6	25.2
Chapathi	30 g	1 piece	120	3	2	22
Poori	20 g	1	85	1.3	5.2	8.1
Bread	18 g	1 slice	44	1.4	0.2	9
Bun	54 g	1	173	3.9	3.9	30.4
Idli	40 g	1	47	1.6	0.1	10
Dosa plain	50 g	1	108	2	4.8	14.1
Pongal	150 g	1 Katori	202	5.5	6	30.9
Upma	130 g	1 Katori	186	4.2	7.8	24.7
Wheat dosa	50 g	1	131	1.8	7.5	14
Ragi ball	260 g	1	400	8	4	81
Kichidi	150 g	1 Katori	250	6.4	8.5	36.9

(continued)

Food	Quantity	Portion/ size	Energy (kcal)	Protein (g)	Fats (g)	Carbohydrates (g)
Jowar roti	80 g	1	134	4	0.47	25
Rice kanji	190 mL	1 Katori	60	1.3	0.2	13
Rava idli	70 g		108	3.2	3.4	14.6
Tamarind rice	150 g	1 Katori	300	8	11	43
Ragi porridge	190 mL	1 Katori	178	4.7	4.3	30.4
Beaten rice	30 g	2 big spoons	90	0.8	2.3	16.4
Raddish sambar	30 g	1 Katori	104	4.2	3.6	13.6
Thick dal	130 g	1 Katori	148	8.5	2.7	22.1
Rasam	200 mL	1 Katori	30	1.4	1	3.8
Daal	130 mL	1 Katori	139	6.1	4.6	17.9
Tomato chutney	18 g	1 tsp	23	0.6	1.2	2.4
Coconut chutney	18 g	1 tsp	47	0.8	3.8	2.3
Mutton fry	140 g	1 Katori	366	25	25	10
Chicken curry	180 mL	1 Katori	376	16.9	32	5.8
Fish fry	30 g	1 piece	70	7	4.9	0.4
Omelette	78 g	1	154	11.5	11.5	1
Fried egg	46 g	1	113	10.7	7.5	0.5
Milk biscuit	13 g	1	52	11	1.8	8.1
Butter	5 g	1 tsp	36	0	4	0
Ghee	5 g	1 tsp	45	0	20	0
Apple	60 g	1	38	0.2	0.3	9
Banana	60 g	1	93	1	0.2	21.7
Guava	100 g	1	66	1.5	0.2	14.5
Mango	100 g	1	74	0.6	0.4	16.9
Papaya	80 g	1	26	0.5	0.1	5.8
Orange	100 g	1	54	1	0.2	12.1
Sugar	5 g	1 tsp	20	0	0	5

(continued)

Food	Quantity	Portion/ size	Energy (kcal)	Protein (g)	Fats (g)	Carbohydrates (g)
Coffee	120 mL	1 small glass	52	1.2	1.1	9.1
Tea	120 mL	1 small glass	43.2	0.5	0.5	9
Curd rice	120 mL	1 small glass	72	3.8	4.8	4
Milk biscuit	120 mL	1 small glass	80	3.8	4.9	5.3
Butter milk	120 mL	1 small glass	18	1	1.3	0.6
Coconut water	120 mL	1 small glass	29	1.7	0.1	5.3

Example: If we want to calculate the content of energy in 1 katori of curd rice,

$$\text{Protein content} = 4.5\text{ g} = 4.5 \times 4\text{ kcal/g} = 18\text{ kcal}$$

$$\text{Carbohydrate} = 25.2\text{ g} = 25.2 \times 4\text{ kcal} = 100.8\text{ kcal}$$

$$\text{Fat} = 3.6\text{ g} = 3.6 \times 9\text{ kcal/g} = 32.4\text{ kcal}$$

$$\begin{aligned}\text{Total energy content} &= \text{Protein content} + \text{Carbohydrate content} + \text{Fat content}\\ &= 18\text{ kcal} + 100.8\text{ kcal} + 32.4\text{ kcal}\\ &= 151.2\text{ kcal}\end{aligned}$$

Glycemic Index

The glycemic index (GI) of a food is defined as the area under the two-hour blood glucose response curve (AUC) following the ingestion of a fixed portion of carbohydrates, usually 50 g.

$$\text{Glycemic index} = \frac{\text{Glucose response to test food}}{\text{Glucose response to standard food}} \times 100$$

The amount of food is standardised to 50 g of carbohydrates. Glucose response is the area under the curve of blood glucose measurements taken 3 hours after the meal.

Standard food is either 50 g of glucose or white bread. Glycemic index is:

- An indicator of the ability of different types of foods that contain carbohydrate to raise the glucose levels within 2 hrs.
- A measure of effects of carbohydrates on blood sugar levels.

Low glycemic foods

Food items which break down quickly during digestion releasing glucose rapidly into circulation have a high glycemic index. For example, oats, milk, carrot, lentils and kidney beans.

High glycemic index foods

Food items that break down slowly, releasing glucose gradually into the blood stream. For example, white rice, cereals, white bread, pumpkin and French fries.

Importance of Glycemic Index

- Low glycemic foods have been reported to improve serum lipid profile, CRP concentrations and help in weight control.
- Low glycemic foods have been associated with higher levels of high density lipoprotein cholesterol (HDL cholesterol), reduced CRP concentrations and decreased risk of developing diabetes and cardiovascular disease.

30 Saturated, Unsaturated and Trans-Fatty Acids in Foods

BI 11.24

Fatty acids are aliphatic carboxylic acids and have the general formula, R–CO–OH. They may be classified in the following ways:

1. Depending on number of carbon atoms
2. Depending on length of hydrocarbon chain
3. Depending on nature of hydrocarbon chain
   - Saturated
   - Unsaturated (monounsaturated and polyunsaturated)

Saturated Fatty Acids

These fatty acids do not contain double bonds in the hydrocarbon chain. They are in extended confirmation and can be packed together in compact structure. They are uniquely resistant to heat and rancidity and are essential for proper nutrient absorption.

Excess of saturated fats increase LDL and decrease HDL cholesterol.
Examples: Butyric acid in butter, lauric acid in coconut oil and myristic acid in cow's milk.

Unsaturated Fatty Acids

These fatty acids contain double bonds in their hydrocarbon chains. Fatty acids with one double bond are **monounsaturated** and those with 2 or more double bonds are collectively known as **polyunsaturated** fatty acids (**PUFA**). They are in cis configuration. This configuration produces a rigid bend in the aliphatic chain, preventing compact packaging and consequently, the unsaturated fatty acids are loosely packed and therefore more easily disturbed by heat. This accounts for the low melting point of unsaturated fatty acids.

PUFA with a double bond between C–3 and C–4 are called omega-3 fatty acids.
Examples: Linolenic acid, eicosapentaenoic acid (EPA), and (DHA)

PUFA with a double bond between C-6 and C-7 are called omega-6 fatty acids.
Examples: Linoleic acid and arachidonic acids

Omega-3 fatty acids are required for brain development and are hence included in the infant formulas. Omega-6 fatty acids are cardiac protective.

Biomedical Importance of PUFA

PUFAs are seen in vegetable oils. They are structural constituents of tissues, gonads and mitochondrial membrane. PUFAs help in the synthesis of prostaglandins and other compounds.

Fats with high content of PUFA tend to lower serum cholesterol. They are essential for optimal vision and immune function. Ingestion of fats rich in PUFA increases fibrinolytic activity and help in skin protection. PUFA are, however, liable to undergo peroxidation and membranes containing PUFA are more prone to damage by free radicals.

Trans-fats

Unsaturated fatty acids can occur in two isomeric forms: cis and trans. Trans-fats are unsaturated fatty acids with trans configuration. There are two types of trans fats in foods—naturally occurring and artificial trans-fats. Trans-fatty acids are present in high amounts in processed foods, fast foods, bakery items and fried food. They compete with essential fatty acids and hence, aggravate essential fatty acid deficiency. They raise LDL levels and lower HDL. Consumption of trans fats on long term may increase the risk of cardiovascular diseases and diabetes. Artificial trans-fats are formed during partial hydrogenation of vegetable oils. They are widely used in food industry because of their long shelf life.

The 2021 guidelines of the American Heart Association recommend liquid plant oils rather than tropical oils (coconut, palm, and palm kernel), animal fats (butter and lard), and partially hydrogenated fats. Saturated and trans fats (animal and dairy fats, and partially hydrogenated fat) should be replaced with non-tropical liquid plant oils to reduce cardiovascular diseases and risk of stroke.

Appendix I

Practical Biochemistry Question Bank

Qualitative Experiments

Essay Questions

1. Analyse the given sample of urine and identify the abnormal constituents present in it.
2. How do you analyse a specimen of urine? Write an essay on the abnormal constituents of urine, stating their diagnostic importance.

Short Notes and Brief Answer Questions

1. Why do you call Benedict's reaction 'semi-quantitative'? How do you interpret it?
2. Jaffe's reaction
3. Composition of normal urine
4. How do you measure the specific gravity of urine?
5. Rothera's test and its clinical significance
6. Fouchet's test and its clinical significance
7. Proteinuria

Quantitative Experiments

Essay Questions

1. Describe in detail the GOD–POD method for glucose estimation. Add a note on the clinical significance of glucose estimation in blood.
2. What is GTT? Describe the procedure. How do you interpret the results of GTT?
3. Describe in detail the methodology adopted to estimate urea levels in blood and clinically interpret the different values.
4. How do you estimate creatinine concentration in serum/urine. Add a note on the clinical significance of creatinine levels.
5. How do you estimate serum uric levels and what is its clinical application?
6. How do you estimate total serum proteins by the biuret method? Add a note on the clinical significance of the A/G ratio.
7. How do you measure the total cholesterol levels in serum by Zak's method? Add a note on the clinical significance of serum cholesterol levels.
8. What is the Van den Bergh reaction? How do you measure serum bilirubin levels by the Malloy and Evelyn method. Add a note on the clinical significance of hyperbilirubinemia.
9. What are aminotransferases? How do you estimate their serum levels? What is their diagnostic significance?
10. Describe the diagnostic significance of ALP and ACP. What are the salient features of their estimation in serum?
11. What are the indications for the estimation of CK? Write a brief note on CK measurement in serum.

12. What is the diagnostic importance of serum amylase and lipase? How do you measure them in serum?
13. What is the diagnostic importance of measuring serum calcium and phosphate levels?
14. How do you analyse CSF? Add a note on its diagnostic application.
15. What is the principle involved in electrophoresis? Write a note on the role of electrophoresis in the diagnosis of different conditions.
16. What is chromatography? Describe the method of paper chromatography. What are the biological applications of chromatography?
17. Explain the principles and applications of radioimmunoassay (RIA) and enzyme-linked immunosorbant assay (ELISA) techniques.
18. Describe how pH is estimated and explain the preparation of different types of buffers.
19. Describe the steps of biomedical waste management and add a note on the safety measures to be adopted.
20. Describe how isolation of DNA is achieved and how it is quantified.
21. Discuss the basis and rationale of the biochemical tests done in the diagnosis and monitoring of diabetes mellitus.
22. Describe briefly the biochemical investigations done and the clinical interpretation of the same in the case of the following conditions:
(i) Dyslipidemia; (ii) myocardial infarction; (iii) jaundice; (iv) acid–base disorders; and (v) thyroid disorders.

Short Notes and Brief-answer Questions

1. Beer–Lambert's law
2. Anticoagulants
3. Quality control
4. FBS and PLBS
5. Impaired glucose tolerance (IGT)
6. Renal glycosuria
7. Alimentary glycosuria
8. Blood urea nitrogen (BUN)
9. DAM method for blood urea estimation
10. Blood urea levels and their clinical interpretation
11. Serum and urinary creatinine and its clinical significance
12. Creatinine coefficient and creatinine clearance
13. Gout and hyperuricemia
14. A/G ratio
15. Serum cholesterol levels – clinical relevance
16. Van den Bergh reaction
17. AST and ALT – diagnostic application
18. Diagnostic significance of ALP and ACP
19. CK – diagnostic importance
20. Amylase and lipase
21. Diagnostic importance of serum calcium and phosphorus
22. CSF analysis
23. Electrophoresis
24. Chromatography

Appendix II Multiple-choice Questions (MCQs) in Practical Biochemistry

1. **The specific test to identify creatinine is**
   (a) sodium hypobromite
   (b) phosphotungstic acid reduction test
   (c) biuret test
   (d) Jaffe's test

2. **Gout is characterised by**
   (a) increased urea levels
   (b) increased uric acid levels
   (c) decreased uric acid levels
   (d) increased creatinine levels

3. **Murexide test is specific for**
   (a) urea (b) creatinine (c) uric acid (d) none of the above

4. **Urinometer is used to measure the**
   (a) specific gravity of urine
   (b) viscosity of blood
   (c) osmotic pressure of blood
   (d) none of the above

5. **Specific gravity of urine is low in**
   (a) diabetes mellitus
   (b) diabetes insipidus
   (c) chronic renal failure
   (d) all of the above

6. **Absence of urobilinogen in urine is suggestive of**
   (a) hemolytic jaundice
   (b) hepatocellular jaundice
   (c) obstructive jaundice
   (d) all of the above

7. **Tubular maximum for glucose (TMG) is**
   (a) 100 mg/min (b) 200 mg/min (c) 500 mg/min (d) 350 mg/min

8. **The following test is used to detect the presence of sugar in urine (glycosuria)**
   (a) biuret test (b) Benedict's test (c) Molisch's test (d) all of the above

9. **Ketonuria is seen in**
   (a) severe uncontrolled diabetes mellitus
   (b) acute renal failure
   (c) cirrhosis of the liver
   (d) all of the above

10. **Rothera's test is used to detect the following substance in urine**
    (a) glucose (b) ketone bodies (c) urobilinogen (d) calcium

11. **Microalbuminuria is considered a risk factor for**
    (a) kidney disease (b) liver disease (c) muscle disease (d) lung disease

12. **Bence–Jones proteins in urine samples are diagnostic of**
    (a) CRF
    (b) DM
    (c) prostate cancer
    (d) multiple myeloma

13. A positive Hay's test is indicative of
(a) calcium in urine (b) bile salts in urine
(c) potassium in urine (d) urobilinogen in urine

14. Pettenkoffer's test is specific for
(a) bile pigments (b) proteins (c) bile salts (d) blood

15. A positive Fouchet's test is indicative of
(a) proteins (b) glucose (c) bile salts (d) bile pigments

16. Concentration of substances in samples are determined by measuring the light absorption at different wavelengths. The amount of light absorbed depends on the following
(a) Beer's law (b) Lambert's law (c) both (d) none

17. Logarithmic ratio of incident light (Io) to that of transmitted light (IE) is known as
(a) optical density (b) transmittance
(c) absorbance (d) percentage transmittance

18. In autoanalysers, the following type of analysis is used
(a) end point analysis (b) kinetic analysis (c) both (d) none

19. The following instrument is used to estimate the concentrations of electrolytes
(a) colorimeter (b) spectrophotometer
(c) ABG analyser (d) flame photometer

20. The following inhibits glycolysis when mixed with a sample of whole blood
(a) potassium oxalate (b) heparin
(c) sodium fluoride (d) none of the above

21. One of the following is an enzymatic method used in the estimation of blood glucose
(a) Folin–Wu method (b) ferricyanide method
(c) Somogyi–Nelson method (d) glucose oxidase method

22. One of the following substances is known to interfere with blood glucose estimation
(a) uric acid (b) glutathione (c) ascorbic acid (d) all of the above

23. Normal postprandial blood glucose concentration should be
(a) less than 140 mg/dL (b) less than 100 mgs/dL
(c) less than 200 mg/dL (d) more than 200 mg/dL

24. GTT is indicated in all of the following categories except
(a) for follow-up in a known case of diabetes
(b) in persons with symptoms and signs suggestive of diabetes but the blood sugar values are inconclusive
(c) in unexplained glycosuria
(d) in diagnosis of GDM

25. A fasting sample in GTT showing plasma glucose concentration 120 mg/dL and two-hour post-load sample of 190 mg/dL with two-hour post-load urine specimen giving positive Benedict's test indicates
(a) normal response (b) DM (c) IGT (d) renal glycosuria

26. In certain persons, despite blood glucose concentration being normal, glucose appears in urine with low renal threshold. Such a condition is known as
(a) IGT (b) alimentary glycosuria
(c) diabetes (d) renal glycosuria

27. The term GDM (gestational diabetes mellitus) represents
(a) diabetic women who become pregnant
(b) hyperglycemia identified for the first time during pregnancy
(c) pregnant women with blood glucose concentrations falling in the IGT range
(d) none of the above

28. Urea is the end metabolite of
(a) carbohydrate metabolism (b) lipid metabolism
(c) amino acid metabolism (d) nucleotide metabolism

29. Urea levels in blood are elevated in
(a) dehydration (b) acute glomerulonephritis
(c) stones in the urinary tract (d) all of the above

30. The following metabolite is a better indicator of kidney function
(a) urea (b) creatinine (c) uric acid (d) all of the above

31. Creatinuria is observed in
(a) HTN (b) brain disease
(c) lung disease (d) muscular dystrophies

32. The major part of uric acid is excreted through
(a) kidneys (b) lungs (c) skin (d) all of the above

33. The most frequently affected joint in gouty arthritis is
(a) knee joint (b) elbow joint
(c) shoulder joint (d) metatarsophalangeal joint

34. The drug of choice to treat hyperuricemia is
(a) allopurinol (b) ibuprofen (c) paracetamol (d) steroids

35. A/G ratio is reversed in the following condition
(a) cirrhosis of the liver (b) nephrotic syndrome
(c) multiple myeloma (d) all of the above

36. The normal serum cholesterol levels in a healthy person should be
(a) 50–100 mg/dL (b) 150–200 mg/dL
(c) 250–300 mg/dL (d) 300–350 mg/dL

37. Elevated levels of only conjugated bilirubin are observed in
(a) hemolytic jaundice (b) hepatocellular jaundice
(c) obstructive jaundice (d) all of the above

38. The levels of the following enzyme are elevated in serum during myocardial infarction
(a) ALT (b) AST (c) ALP (d) ACP

39. The concentration of the following enzyme is markedly elevated in hepatocellular jaundice
(a) ALT (b) AST (c) both (d) none

40. Obstructive jaundice is characterised by marked elevation of serum levels of
(a) ALT (b) AST (c) ACP (d) ALP

41. Acid phosphatase levels in serum are elevated in
(a) Paget's disease (b) prostate cancer
(c) obstructive jaundice (d) none of the above

42. The levels of the following enzyme are elevated in acute myocardial infarction
(a) LDH (b) AST (c) CK (d) all of the above

43. Serum amylase levels are elevated in
(a) acute pancreatitis (b) acute gastritis
(c) carcinoma lung (d) prostate cancer

44. The normal concentration of Ca^{2+} in plasma
(a) 15–20 mg/dL (b) 20–25 mg/dL (c) 9–11 mg/dL (d) 6–8 mg/dL

45. Pandy's test is used to measure the following substance in CSF
(a) glucose (b) chlorides (c) lipids (d) globulins

46. The following plasma protein moves faster during electrophoretic separation
(a) albumin (b) α-globulin (c) β-globulin (d) γ-globulin

47. M band is seen in
(a) normal individuals (b) liver cirrhosis
(c) nephrotic syndrome (d) multiple myeloma

48. The following physicochemical principle is involved in chromatographic separation
(a) partition (b) ion exchange (c) affinity (d) all of the above

49. The ratio of the distance moved by a compound to the distance moved by the solvent front is referred to as
(a) Rs value (b) Rp value (c) Rf value (d) none of the above

50. The following biomolecule is separated by chromatography
(a) sugars (b) lipoproteins (c) amino acids (d) all of the above

Answers

1. d	2. b	3. c	4. a	5. b	6. c	7. d	8. b	9. a	10. b
11. a	12. d	13. b	14. c	15. d	16. c	17. a	18. b	19. d	20. c
21. d	22. d	23. a	24. a	25. c	26. d	27. b	28. c	29. d	30. b
31. d	32. a	33. d	34. a	35. d	36. b	37. c	38. b	39. c	40. d
41. b	42. d	43. a	44. c	45. d	46. a	47. d	48. d	49. c	50. d

Appendix III Problem-solving Exercises, Interpretation of Laboratory Data and Case Studies in Practical Biochemistry

1. **Problem: Calculate in mmol/L, a 90 mg/dL concentration of glucose.**

Ans. Molecular formula of glucose: $C_6H_{12}O_6$

Molecular weight of $C_6H_{12}O_6$ = 180

$$\text{mmol/L} = \frac{\text{Mass in mg}}{\text{Molecular weight}}$$

$$= \frac{90}{180} \times 10 = 5 \text{ mmol/L}$$

2. **Problem: Express in mEq/L the serum sodium concentration of 322 mg/dL.**

Ans. $$\text{mEq/L} = \frac{\text{mg/dL} \times \text{Valency}}{\text{Molecular weight}} \times 10$$

$$= \frac{322 \times 1}{23} \times 10 = 140 \text{ mEq/L}$$

3. **Problem: How will you prepare 250 mL of 0.1 M HCl from a stock of 1 M HCl?**

Ans. Volume of HCl = 250 mL

Dilution = 1 in 10 dilution

Concentration of stock standard = 1 Molar

Concentration of working standard = 0.1 M

$V_1 \times C_1 = V_2 \times C_2$

$V_1 \times 1\text{ M} = 250\text{ mL} \times 0.1\text{ M}$

$V_1 = 25\text{ mL}$

Thus, take 25 mL of stock and dilute with distilled water to make up to 250 mL.

4. **Problem: What is the degree of dilution when a working standard of 0.2 mg/mL is required to be prepared from a stock of 1 g/dL?**

Ans. Stock concentration (C_1) = 100 mg/dL

Stock volume (V_1) = 100 mL

Working standard concentration (C_2) = 0.2 mg

Working standard volume (V_2) = 1 mL

$$\frac{V_1}{V_2} = \frac{C_1}{C_2} = \frac{100}{1} = \frac{1000}{0.2} = 1:50 \text{ dilution is required.}$$

5. **Case Study**

A 45-year-old man comes to clinic complaining of increased appetite and thirst and more frequent urination. Upon testing his blood and urine, the following results are obtained.

	Fasting	Post-lunch
Blood sugar	160 mg/dL	220 mg/dL
Urine sugar	–ve	+ve

Interpretation

The patient presents with classic symptoms of diabetes and his blood glucose levels and glycosuria are the supporting evidence of hyperglycemia. He can be diagnosed as suffering from **diabetes mellitus**.

6. **Identify and interpret the following chart.**

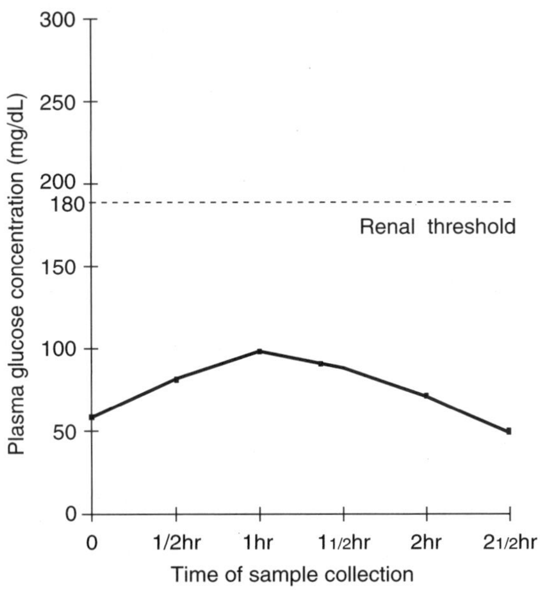

Urine specimen						
Benedict's test	–ve	–ve	–ve	–ve	–ve	–ve
Rothera's test	–ve	–ve	–ve	–ve	–ve	–ve

Interpretation

Glucose concentration in all the six samples estimated is within the normal range. No glycosuria is detected. Hence the curve depicted above represents **normal GTT**.

7. **Identify and interpret the following graphical chart.**

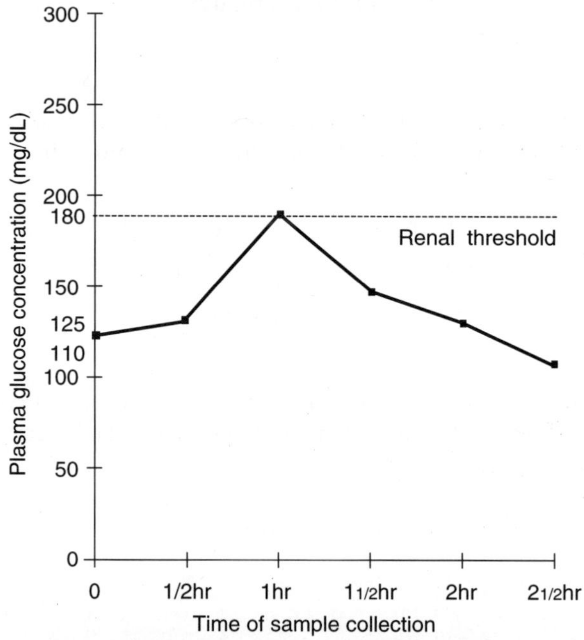

Urine specimen						
Benedict's test	–ve	–ve	+ve	–ve	–ve	–ve
Rothera's test	–ve	–ve	–ve	–ve	–ve	–ve

Interpretation

Note that in the above example the plasma glucose levels are above normal but below diabetic levels. These values represent an intermediary stage known as **impaired glucose tolerance (IGT)**. Urine samples tested for both glucose and ketone bodies are negative. Benedict's test may be positive in one or two samples if the glucose level rises beyond 180 mg/dL in post-load samples.

These persons are normal but at an increased risk of developing diabetes. Hence a change in lifestyle is advised (diet restriction and exercise).

	Fasting	Post-load
Plasma glucose levels in IGT	110–125 mg/dL	140–199 mg/dL
Urine specimen	–ve	–ve/+ve

8. **Interpret the following chart**

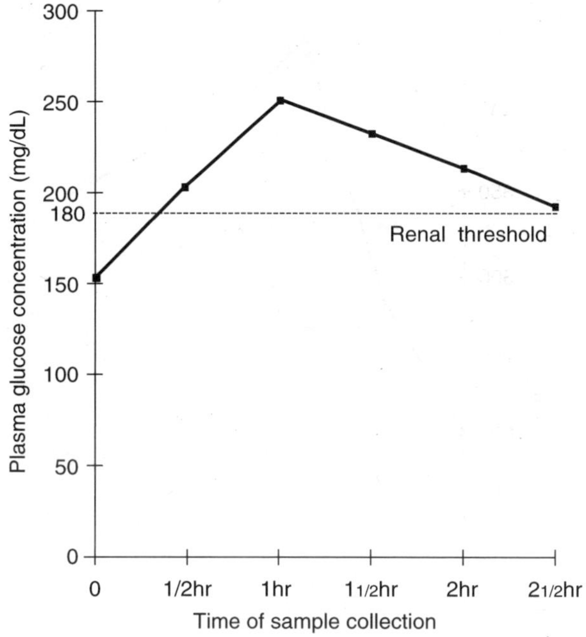

Benedict's test	–ve	–ve	+ve	+ve	+ve	+ve
Rothera's test	–ve	–ve	–ve	–ve	–ve	–ve

Interpretation

Glucose tolerance is always reduced in diabetes mellitus. The plasma glucose concentrations are above the normal levels. Benedict's test is positive for post-load urine samples. Note that the high glucose values are persistent even after two hours. This is a case of **moderate diabetes** (note the absence of ketone bodies in urine).

	Fasting	Post-load
Plasma glucose concentration in diabetes	≥ 126 mg/dL	≥ 200 mg/dL

9. **Study the following graph and interpret it.**

Urine specimen

Benedict's test	+ve	+ve	+ve	+ve	+ve	+ve
Rothera's test	+ve	+ve	+ve	+ve	+ve	+ve

Interpretation

Note the very high levels of plasma glucose. All urine samples are positive for Benedict's test. Ketone bodies are detected in urine specimens (positive Rothera's test). The graph suggests severe **uncontrolled diabetes millitus.**

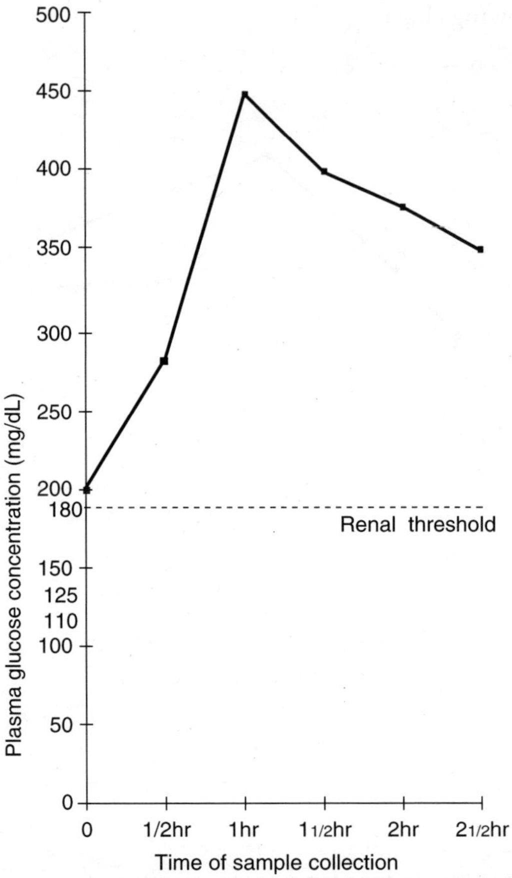

10. Interpret the following chart

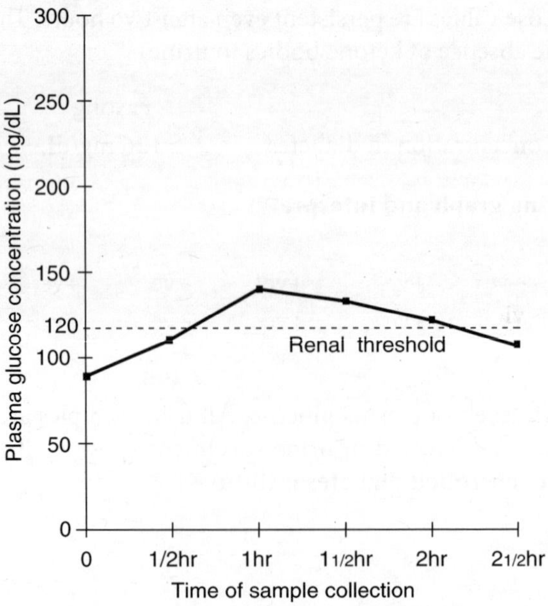

Urine specimen						
Benedict's test	+ve	−ve	+ve	+ve	+ve	+ve

Interpretation

The plasma glucose concentrations in the above example are within the normal range. Despite the normal concentrations in blood, glucose appears in urine (positive Benedict's test). This is due to lowering of the renal threshold. Such a condition is known as **renal glycosuria**. It is a harmless condition and requires no specific treatment. Renal glycosuria is commonly seen during pregnancy.

11. Identify the following graph.

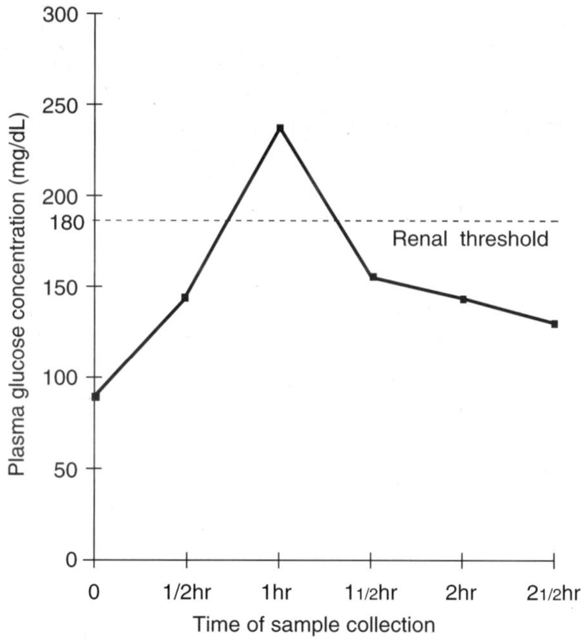

Urine specimen						
Benedict's test	−ve	−ve	+ve	−ve	−ve	−ve

Interpretation

In the above graph, the fasting and two-hour glucose values are normal, but a sudden rise in glucose level following the ingestion of glucose is noticed. This is due to the very high rate of absorption of glucose from the intestine. The one-hour urine specimen shows glycosuria. The condition is known as **alimentary glycosuria**. It is a normal condition seen in some individuals. It may be due to diminished carbohydrate intake for a few days.

12. Case Study

In a patient, the blood urea concentration is 80 mg/dL and urinary urea 3200 mg/dL. Rate of urine flow is 1 mL/minute. Calculate the urea clearance and interpret the result.

Ans. Urea clearance $= \frac{U \times V}{P} = \frac{3200 \times 1}{80}$

$= 40$ mL/minute

The normal urea clearance is about 75 mL/minute.

The obtained value (40 mL/minute) indicates that there is approximately 50% deficit of kidney function in the patient.

13. Why is the creatinine clearance test preferred over estimation of serum concentration in elderly patients?

Ans. In the elderly, the skeletal muscle mass is grossly reduced, and as a result, serum levels of creatinine are also reduced to such an extent that they may be in the normal range even in severe renal disease.

14. Urea is normal and creatinine is high in patients of diabetic ketoacidosis. Explain.

Ans. Jaffe's reaction, which is used to estimate creatinine, is not specific for creatinine. Certain other substances such as ketone bodies, proteins, glucose and pyruvate react with picric acid and give false measurements. In normal individuals, the false high levels may be up to 10–15 percent of 'true' creatinine concentration. In diabetic ketoacidosis, excess ketone bodies react with picric acid, giving high false values of creatinine concentration.

15. Case Study

An obese 55-year-old man working as an executive in an MNC attends luxurious parties regularly. One evening, he attended a lavish dinner with drinks, went home and slept. All of a sudden, he was awoken from sleep by excruciating pain in his right big toe and was unable to put his foot on the floor.

On examination, the doctor found the joint of the toe swollen, red and very tender. He was treated with indomethacin and the symptoms subsided completely. On enquiry, he gave a history of a previous attack of renal colic a year back and considered himself too busy to be investigated further.

The doctor sent his serum sample to the laboratory for estimation of uric acid and found very high levels (20 mg/dL).

What is your interpretation of this case?

Interpretation

This is a classic case of **gout**. The sudden onset, at night, is common. In most cases of gout, the joint affected is the metatarsophalangeal joint of the big toe. The classic signs of inflammation (swelling, redness, tenderness, etc.) were present and upon investigation, hyperuricemia was confirmed. In this case, the history of renal colic could be an outcome of a renal urate stone. Further, gout is more common in men than in women and is associated with high living, alcohol consumption and obesity.

16. Case Study

A four-year-old girl from a low socio-economic group family presents with edema, muscle wasting and paleness (pallor). On examination, growth retardation, delayed milestones and anemia were found. What investigations would you suggest and what is your probable diagnosis?

Interpretation

It is a classic case of protein malnutrition (**kwashiorkor**). Serum total protein levels are very low. Low albumin levels results in edema all over the body.

17. Case Study

A five-year-old boy presents with edema on the face and anasarca (generalised edema all over the body). On investigation, total serum protein was around 3 mg/dL and serum cholesterol was as high as 650 mg/dL. Urine was characterised by proteinuria.

What is your probable diagnosis?

Interpretation

Increased excretion of proteins in urine decreases plasma protein concentration, which in turn imbalances the 'starling forces' (osmotic pressure, hydrostatic pressure, etc.), resulting in abnormal accumulation of fluid in tissue spaces (edema). This is a case of **nephrotic syndrome**. In cases of nephrosis, usually the cholesterol levels are very high.

18. Case Study

A 45-year-old woman consulted her physician with a complaint of abdominal pain which appeared on and off. The pain was mostly localised in the upper abdomen, sometimes radiating to the chest. The distress usually followed the intake of food, especially after consuming fried items. She experienced severe nausea and vomiting during the acute episodes. No history of previous attacks of jaundice or GI bleeding was reported. Serum cholesterol levels were high.

What is your probable diagnosis and advice to the patient?

Interpretation

It could be 'irritable bowel syndrome' and the patient is given antacids. If there is no relief, consider the diagnosis of **cholestasis (gall stones)**. If the diagnosis of cholestasis is confirmed, the patient is advised to undergo cholecystectomy.

19. Case Study

A newborn develops yellowish discolouration of the skin and conjunctiva three days after birth. The condition is handled by giving phototherapy.

What type of jaundice is it and name the biochemical tests which help in its diagnosis.

Interpretation

In a newborn (especially when it is premature), the liver is not fully developed, resulting in jaundice perhaps due to failure to conjugate bilirubin. This type of jaundice is known as **neonatal physiological jaundice**. To confirm the diagnosis, serum levels of bilirubin, ALT, AST and urine for bile pigments should be tested.

20. Interpretation of laboratory data

The following are some of the findings of biochemical investigations in a patient.

Van den Berg reaction	–	Indirect positive
Total bilirubin	–	11.4 mg/dL
Conjugated bilirubin	–	0.6 mg/dL
Unconjugated bilirubin	–	10.8 mg/dL

What is your probable diagnosis?

Ans. Hemolytic jaundice

21. Interpretation of laboratory data

The following are some of the findings of biochemical tests in a patient.

Van den Bergh reaction	–	Biphasic
Total bilirubin	–	12.8 mg/dL
Conjugated bilirubin	–	6.4 mg/dL
Unconjugated bilirubin	–	6.4 mg/dL
ALT	–	286 IU/L
AST	–	264 IU/L

What is your probable diagnosis?

Ans. Hepatocellular jaundice

22. Interpretation of laboratory data

The following are some of the biochemical findings in a patient.

Van den Bergh reaction	–	Direct positive
Total bilirubin	–	15.6 mg/dL
Conjugated bilirubin	–	14.2 mg/dL
Unconjugated bilirubin	–	1.4 mg/dL
ALT	–	38 IU/L
AST	–	20 IU/L
ALP	–	450 IU/L

What is your probable diagnosis?

Ans. Obstructive jaundice

23. Interpretation of laboratory data

The following are the biochemical findings in a patient brought to the hospital in coma.

Plasma random sugar	–	460 mg/dL
Benedict's test with urine	–	brick red
Rothera's test	–	positive
Serum HCO_3^-	–	16 mEq/L
Plasma pH	–	7.25

What is your probable diagnosis?

Ans. Diabetic ketoacidosis

24. Interpretation of laboratory data

Two patients were admitted to the hospital with the complaint of chest pain. The following investigations were performed in each of them.

1st patient	2nd patient
CK – 2968 IU/L	CK – 76 IU/L
AST – 760 IU/L	AST – 34 IU/L
LDH – 1120 IU/L	LDH – 170 IU/L

What is your diagnosis?

Ans. 1st case – Myocardial infarction
2nd case – Simple myalgia

25. Interpretation of laboratory data

A 50-year-old man was admitted in the hospital with acute abdominal pain. Serum amylase estimation gave a value of 430 IU/L.

What is the probable diagnosis?

Ans. Acute pancreatitis

26. Interpretation of laboratory data.

An adult male was brought to the hospital with complaints of pain and muscle cramps following exercise. He gave a history of myoglobinuria. Upon investigation, creatine kinase levels were found to be very high (1200 IU/L).

What is your probable diagnosis?

Ans. Muscular dystrophy

27. Interpret the following electrophoretic slide.

Ans. Normal pattern

Interpretation

Normal serum separates into five bands when subjected to electrophoretic separation- albumin, α_1, α_2, β and γ fraction of globulins. Albumin is maximally mobile due to its relatively low molecular weight and γ-globulins are minimally mobile. α_1 fraction is mostly composed of α_1-antitrypsin, α_2-fraction is mostly made up of α-2-macroglobulin. β-fraction contains mostly LDL (lipoproteins).

If plasma is used instead of serum for electrophoresis, the fibrinogen will form a band in the γ-globulin region.

28. Interpret the following electrophoretic separation of plasma proteins.

Ans. Cirhosis of the liver

Interpretation

Note the decreased albumin and increased β and γ fractions of globulins. Sometimes, the β and γ fractions are fused. In cirrhosis of the liver, albumin synthesis by the liver is diminished. To maintain the normal total protein concentration, a compensatory excess synthesis of globulins by the reticulo-endothelial system occurs.

29. Interpret the following electrophoretic separation of plasma proteins.

Ans. Nephrotic syndrome

Interpretation

Observe the diminished albumin and prominent α-2 band of globulins.

In nephrotic syndrome, all plasma proteins are lost in urine. α-2-globulin fraction, composed mostly of α-2-macroglobulin (which is the biggest), escapes this and is retained showing a prominent band.

30. Interpret the following electrophoretic slide

Ans. Multiple myeloma

Interpretation

Observe a new band (M-band) between the β and γ fractions of globulins. The M band ('M' stands for monoclonal) in multiple myeloma (a plasma cell cancer) is due to the para proteinemias.

31. Case Study

A 60-year-old man was brought to hospital in a semiconscious state. The patient displayed a typical hyperventilatory breathing pattern with a sweetish odour of breath. The pulse was feeble and the BP was low. Blood gas analysis and other laboratory investigations were performed and the results obtained were as follows.

Parameter	Obtained value	Reference range
Blood pH	7.1	7.35–7.45
Plasma HCO_3^-	11 mmol/L	20–30 mmol/L
pCO_2	38 mmHg	35–45 mmHg
Blood glucose (random)	550 mg/dL	less than 140 mg/dL

Benedict and Roth's tests were positive in the urine samples. How would you interpret this case and what is your probable diagnosis?

Ans: It is a case of severe diabetes mellitus left untreated resulting in diabetic ketoacidosis. It is a classic example of ***metabolic acidosis***. The blood glucose levels are very high and this is associated with urinary excretion of glucose and ketone bodies. The patient has typical rhythmic respiration (*Kussmaul's respiration*). The fruity odour of the breath is due to the presence of acetone in the exhaled air. Note that pCO_2 values are within normal limits as it is compensated by hyperventilation. The plasma concentration of HCO_3^- is markedly low. This is the typical finding of non-respiratory acidosis. Bicarbonate levels are depleted as they have been consumed to buffer H^+ ions. The reduced HCO_3^- concentration is the primary abnormality in metabolic acidosis and the plasma HCO_3^- levels reflect the degree of severity of the metabolic acidosis.

32. Case Study

A 50-year-old man was admitted to hospital with a complaint of persistent vomiting. On examination he was found to be dehydrated and his respiration shallow. He gave a past history of dyspepsia. The results of the laboratory investigations are as follows.

Parameter	Obtained value	Reference range
Blood pH	7.7	7.35–7.45
plasma HCO_3^-	45 mmol/L	20–30 mmol/L
pCO_2	60 mm Hg	35–45 mmHg
Na^+	140 mEq/L	136–145 mEq/L
K^+	2.5 mEq /L	3.5–5 mEq/L

How can the interpretation of the above laboratory data aid you in your diagnosis?

Ans: Findings of elevated blood pH and an increase in plasma HCO_3^- levels are characteristic of ***metabolic alkalosis***. The increase in pCO_2 is a result of compensatory hypoventilation leading to the retention of carbon dioxide. Dehydration by fluid loss stimulates Na^+ reabsorption in the kidneys. The patient excretes more H^+ and K^+ ions in exchange for reabsorption of Na^+. This is the reason why the patient excretes acidic urine despite alkalosis.

33. Case Study

A young man was brought to casualty after he had met with a road accident that caused injury to his chest wall. The results of the laboratory investigations were as follows.

Parameter	Obtained value	Reference range
Blood pH	7.2	7.35–7.45
Plasma HCO_3^-	26 mmol /L	20–30 mmol/L
pCO_2	60 mmHg	35–45 mmHg

Interpret the laboratory data and diagnosis.

Ans: The low pH and high pCO_2 values are suggestive of ***respiratory acidosis***. The almost normal HCO_3^- levels indicate that no renal compensation has occurred as yet. As discussed in the textbook, renal compensation needs several days to become fully effective, in contrast to the respiratory compensation in metabolic acid–base disorders which occur in few minutes.

34. Case Study

A medical student volunteered a sample of arterial blood for a hematology experiment in the physiology lab. The tutor took some time to explain the procedure to the class, during which time the student became increasingly anxious and started complaining of a tingling feeling and numbness in his fingers and toes while the blood was being drawn. The results of the investigations of the arterial blood are as follows.

Parameter	Obtained value	Reference range
Blood pH	7.5	7.35–7.45
Plasma HCO_3^-	24 mmol /L	20–30 mmol/L
pCO_2	25 mmHg	35–45 mmHg

What is your comment?

Ans: There is an increase in blood pH and lowered pCO_2 indicating the the student was becoming alkalotic. Going by the reduced pCO_2, it can be labelled as a ***respiratory alkalosis***. The low pCO_2 is a result of anxiety induced hyperventilation, and no compensation has occurred in such a short time span. The symptoms of tingling and numbness are associated with a decline in the calcium ion levels, which is common in alkalosis.

Appendix IV Test Procedures of Qualitative and Quantitative Experiments in a Nutshell

Qualitative Test Procedures (in alphabetical order)

1. **Aldehyde Test**: To 2 mL of the given solution, add 2 drop of dilute formalin and 1–2 drops of 10% mercuric sulphate. Mix well. Add 2 mL of concentrated H_2SO_4 along the side walls of the test tube. Observe the violet colour at the junction of the two liquids. This indicates the presence of **tryptophan.**
2. **Barfoed's Test**: To 4 mL of Barfoed's reagent, add 2 mL of the given solution and heat for about 30–60 seconds, or keep the test tube in a boiling water bath for about 3 minutes. A red scum will appear all over the walls of the test tube and a red precipitate will be seen at the bottom of the test tube upon cooling. This is indicative of a **monosaccharide.**
3. **Benedict's Test**: To 5 mL of Benedict's qualitative reagent, add 0.5 mL (8 to 10 drops) of the given solution and heat for about 2 minutes or keep in a boiling water bath for about 2–5 minutes. A brick red–coloured precipitate will appear, indicative of **reducing sugar.**
4. **Benedict's Test after Acid Hydrolysis**: To 5 mL of the given solution, add 2 drops of concentrated HCl. Boil for 1 minute. Now neutralise the solution with solid 20% sodium carbonate solution till the effervescence ceases. Boil 1 mL of neutralised solution with 5 mL of Benedict's reagent. A brick red–coloured precipitate will appear, indicating the presence of **reducing sugar.**
5. **Benzidine Test**: Mix 2–3 drops of benzidine solution and 2 drops of H_2O_2. Add 2 drops of this mixture to 2 mL of the given solution in a second test tube. A blue or green colour develops, indicating the presence of **blood.**
6. **Biuret Test**: To 2 mL of protein, add 2 mL of 5% NaOH and 2 drops of 1% $CuSO_4$ and mix. A purple or violet colour is indicative of protein. It is the **general reaction for all proteins.**
7. **Ethereal Sulphate Test**: To 5 mL of the given solution, add 2 mL of $BaCl_2$ and 2 mL of concentrated HCl. Mix and filter. Divide the filtrate into two portions. Boil one and compare with the other (control). Turbidity indicates the presence of **ethereal sulphates.**
8. **Fouchet's Test**: To 10 mL of the given solution, add 5 mL of $BaCl_2$ and a pinch of $MgSO_4$. Mix well. $BaSO_4$ is precipitated. After 5 minutes, filter the solution. Unfold the filter paper over a dry filter paper. Add a few drops of Fouchet's reagent on the precipitate. Green or blue colour is indicative of **bile pigments.**

9. **Full Saturation Test**: Take 3 mL of the given solution and add solid ammonium sulphate till the solution is saturated. Let it stand for 5 minutes. Filter. A precipitate is formed. Perform the biuret test with 40% NaOH. A positive test indicates that the protein is not precipitated by full saturation **(peptone).** A negative test indicates that the protein is precipitated by full saturation **(albumin).**
10. **Foulger's Test**: To 3 mL of Foulger's reagent, add 0.5 mL (8–10 drops) of the given solution. Boil for 1 minute and cool. A deep blue colour is indicative of **fructose.**
11. **Gerhardt's Test**: To 3 mL of the given solution, add ferric chloride, drop by drop. A port wine colour indicates the presence of **ketone bodies.**
12. **Gmelin's Test**: To 3 mL of concentrated HNO_3, add 3 mL of the given solution. A green, blue or violet ring indicates the presence of **bile pigments.**
13. **Half-saturation Test**: To 3 mL of the given solution, add 3 mL of saturated ammonium sulphate solution and mix. Allow it to stand for 5 minutes. Filter. A precipitate is formed, with the filtrate (1 mL). Perform biuret test with 40% NaOH. A positive test indicates that the protein is not precipitated by half-saturation **(albumin).** A negative test indicates that the protein is precipitated by half-saturation **(gelatin).**
14. **Heat and Acetic Acid Test or Heat Coagulation Test**: Take 10 mL of the given solution in a test tube. Boil the upper half of the solution and add 1% acetic acid. Coagulum is formed in the heated portion of the test tube, indicating **albumins.**
15. **Hay's Test**: To 2 mL of the given solution, sprinkle a pinch of sulphur powder. Observe without mixing. The sulphur powder sinks to the bottom, indicating the presence of **bile salts.**
16. **Heller's Test**: To 3 mL of the given solution, add 3 mL of concentrated HNO_3 along the sides of the test tube. A white ring formed at the junction of the two liquids indicates protein denaturation by the concentrated acid.
17. **Isoelectric Precipitation**: To 5 mL of the given solution, add 2–3 drops of bromocresol green indicator. A blue colour is seen. Add 1% acetic acid, drop by drop, till a blue-coloured curdy precipitate appears (pH 4.6). A blue-green curdy precipitate develops, indicating **casein.**
18. **Jaffe's Test**: Label two tubes 'test' and 'control'. Take 5 mL of the given solution in 'test' and 5 mL of H_2O in 'control'. To each, add 1 mL of saturated picric acid and 10 drops of 10% NaOH. An orange-red colour indicates the presence of **creatinine.**
19. **Millon's Test**: To 2 mL of the given solution, add 1 mL of 10% mercuric sulphate and boil for 1 minute. Then cool and add 2–3 drops of 1% sodium nitrite. A red colour or precipitate indicates **tyrosine.**
20. **Molisch's Test**: To 2 mL of the given solution, add 2 drops of 1% alcoholic α-naphthol solution and mix. Add 2 mL of concentrated H_2SO_4 along the sides of the test tube to form a separate layer. A purple- or violet-coloured ring is formed at the junction of the two solutions, indicating the presence of carbohydrates. It is the **general reaction for all carbohydrates.**
21. **Murexide Test**: Take 5 mL of the given solution in a china dish, and add 4–5 drops of concentrated HNO_3. Evaporate over a low flame carefully. Allow the dish to cool. Add 1–2 drops of dilute ammonia. A purplish-red colour develops, indicating **uric acid.**
22. **Ninhydrin Test**: To 1 mL of the given solution, add 10 drops of ninhydrin and boil strongly. A purple or blue colour develops, indicating **free amino groups.**

23. **Osazone Test**: Take 1 spatula of phenyl hydrazine HCl and 2 spatula of sodium acetate and 5–10 drops of glacial acetic acid. To this, add 5 mL of the given solution and keep in a water bath for 20 minutes for monosaccharides and 40 minutes for disaccharides. Observe the crystals under a microscope.
Sunflower petal–shaped osazones – **Maltose**
Hedgehog-shaped osazones – **Lactose**
Needle-shaped osazones – **Glucose, fructose**
24. **Phosphorous (Neumann's) Test:** To 5 mL of the given solution, add 1 mL of 40% NaOH. Boil strongly for 1–2 minutes. Cool under tap water. Add 1 mL of concentrated HNO_3. Add a pinch of solid ammonium molybdate and warm gently. A canary-yellow colour develops, indicating **casein.**
25. **Phosphotungstic Acid Test (Benedict's Reduction Test)**: To 2 mL of the given solution, add a few drops of phosphotungstic acid reagent and a few drops of 20% Na_2CO_3. The emergence of a blue colour is indicative of **uric acid.**
26. **Rapid Furfural Test**: To 2 mL of the given solution, add 2–3 drops of 1% alcoholic α-naphthol. Mix and add 3 mL of concentrated HCl. Boil and let it cool. A violet colour is indicative of fructose.
27. **Rothera's Test**: To 10 mL of the given solution, add solid ammonium sulphate, a little at a time, with mixing to saturate the solution. Add 2 or 3 drops of freshly prepared sodium nitroprusside solution. Mix gently, add 1 mL of strong NH_4OH, drop-wise along the sides of the test tube. Do not mix. A purple-coloured ring at the junction of the two liquids indicates **ketone bodies.**
28. **Sakaguchi's Test**: To 2 mL of the given solution, add 3–4 drops of α-naphthol and 6–8 drops of sodium hypobromide. A bright red colour is indicative of **arginine.**
29. **Salicyclic Acid Test**: To 2 mL of the given solution, add a few drops of 20% sulphosalicylic acid. A white-coloured precipitate indicates the precipitation of **proteins.**
30. **Schiff's Test**: Wet a piece of filter paper with a few drops of ammoniacal $AgNO_3$ solution. Add 1 or 2 drops of the given solution on the same paper. A black colour is indicative of **uric acid.**
31. **Seliwanoff's Test**: To 3 mL of Seliwanoff's reagent, add 1 mL of the given solution and boil for 2 minutes and cool. A cherry red colour indicates **fructose.**
32. **Sodium Hypobromite Test**: To 3 mL of the sample solution, add 4–6 drops of sodium hypobromite solution. An effervescence indicates the presence of **urea.**
33. **Specific Urease Test:** Label two test tubes 'test' and 'control'. In 'test', take 5 mL of the given solution and 2 mL of well-mixed urease suspension. In 'control', take 5 mL of the given solution and 2 mL of urease suspension which has been heated strongly. Incubate for 15–20 minutes at room temperature. At the end, add 2 drops of phenolphthalein to both the tubes. A pink colour is indicative of **urea.**
34. **Sulphur Test**: To 2 mL of the given solution, add 2 mL of 40% NaOH and boil strongly and cool. Add 2–3 drops of lead acetate and mix. A black or brown colour is indicative of **cysteine** and **cystine.**
35. **Urobilinogen Test**: To 5 mL of the given solution, add 1 mL of Ehrlich's reagent. Mix. Let it stand for 5 minutes. A red colour can be seen through the mouth of the test tube, indicating **urobilinogen.**
36. **Xanthoproteic Test**: To 3 mL of the given solution, add 1 mL of concentrated HNO_3 and boil for 1 minute. Cool under running tap water and add 40% NaOH till it becomes alkaline.

A yellow colour turns orange on addition of NaOH, indicating the presence of **aromatic amino acids.**

Quantitative Test Procedures

1. Estimation of Blood Glucose (GOD–POD Method)

Reagents (mL)	Blank	Standard	Test
Distilled water	0.05 mL	–	–
Working standard	–	0.05 mL	–
Sample	–	–	0.05 mL
GOD–POD reagent	5.0 mL	5.0 mL	5.0 mL

Mix and keep in an incubator for 15 minutes. and check for reading (540 nm).

OD			

Calculation

$$\text{Concentration of glucose (mg/dL)} = \frac{T}{S} \times 100 \text{ mg/dL}$$

2. Estimation of Blood Urea (DAM Method)

	Blank	Standard	Test
Supernatant	–	–	2.0 mL
Working urea standard	–	2.0 mL	–
Distilled water	2.0 mL	–	–
Diacetyl monoxime reagent	0.4 mL	0.4 mL	0.4 mL
H_2SO_4–H_3PO_4 mixture	1.6 mL	1.6 mL	1.6 mL

Mix well and keep the test tubes in boiling water for 30 minutes. Cool under running tap water. Read at 480 nm (blue filter).

(*Note:* If the colour of the test is stronger than the standard, the test is repeated with a diluted sample, after an appropriate dilution).

	B	S	T
Optical density (OD)			

Calculation

$$\text{Concentration of urea (mg/dL)} = \frac{T}{S} \times 40$$

3. Estimation of Serum Creatinine (Jaffe's Method)

Reagents (mL)	Blank	Standard	Test
Supernatant	–	–	3 mL
Working standard	–	3 mL	–
Distilled water	3 mL	–	–
NaOH (0.75 N)	1 mL	1 mL	1 mL
Picric acid	1 mL	1 mL	1 mL

Mix and keep in an incubator for 15 minutes and check for reading (540 nm).

OD			

Calculation

$$\text{Concentration of creatinine (mg/dL)} = \frac{T}{S} \times 2$$

4. Estimation of Serum Uric Acid (Caraway's Method)

	Blank	Standard	Test
Distilled water	3.0 mL	–	–
Uric acid working standard (0.84 mg/100 mL)	–	3.0 mL	–
Supernatant	–	–	3.0 mL
Sodium carbonate	0.6 mL	0.6 mL	0.6 mL
Phosphotungstate	0.6 mL	0.6 mL	0.6 mL

Mix well and keep them at room temperature for 30 minutes. Take the reading at 700 nm or by using a red filter within 15 minutes after adjusting the blank.

Calculation

$$\text{Concentration of uric acid (mg/dL)} = \frac{T}{S} \times 8.4$$

5. Estimation of Total Serum Protein (Biuret Method)

	Blank	Standard	Test
Distilled water	3.0 mL	–	–
Standard solution	–	3.0 mL	–
Sample	–	–	0.1 mL
Normal saline	–	–	2.9 mL
Biuret working reagent	3.0 mL	3.0 mL	3.0 mL
Optical density (OD) (at 535 nm)			

Calculation

$$\text{Concentration of uric acid} = \frac{T}{S} \times 6 \text{ g/dL}$$

6. Estimation of Serum Cholesterol (Zak's Ferric Chloride Method)

Reagents	Blank	Standard	Test
Cholesterol working standard	–	9.9 mL	–
NaCl (0.9%)	–	0.1 mL	–
Serum	–	–	0.1 mL
$FeCl_3$	–	–	9.9 mL
Mix well and keep for 10 minutes and centrifuge			
Supernatant	–	5 mL	5 mL
$FeCl_3$	5 mL	–	–
H_2SO_4	3 mL	3 mL	3 mL

Mix well and incubate for 30 minutes at room temperature. Read absorbance at 560 nm in colorimeter.

Calculation

$$\text{Concentration of cholesterol} = \frac{T}{S} \times 400 \text{ mg/dL}$$

Spotters and Charts

A Selected List of Spotters

Instruments

1. Colorimeter
2. pH meter
3. Urinometer
4. Ryle's tube
5. Centrifuge
6. Electrophoretic apparatus
7. Chromatographic chamber

Chemicals and Reagents

1. Picric acid
2. Ammonium sulphate
3. Copper sulphate

Chemical Structures

1. Amino acids (any of the 20 standard amino acids)
2. Monosaccharides (Glucose, Fructose, Galactose)
3. Disaccharides (Maltose, Lactose, Sucrose)
4. t-RNA
5. Cholesterol
6. Heme
7. PUFA
   - Linoleic acid
   - Linolenic acid
   - Arachidonic acid

4. Lead acetate
5. Sodium fluoride
6. Benzidine powder
7. Phenyl hydrazine hydrochloride
8. Benedict's qualitative reagent
9. Seliwanoff's reagent
10. Fouchet's reagent

Microscopic Slides

1. Gloucosazone/Fructosazone
2. Maltosazone
3. Lactosazone

A Selected List of Charts

1. **GTT curves**
   - Normal
   - Diabetes mellitus
   - IGT
   - Renal glycosuria
2. **Electrophoretic separation of plasma proteins**
   - Normal
   - Cirrhosis of the liver
   - Nephrotic syndrome
   - Multiple myeloma
3. **Biochemical laboratory data in different types of jaundice**
   - Hemolytic jaundice
   - Hepatocellular jaundice
   - Obstructive jaundice
4. **Biochemical laboratory data in certain diseases**
   - Myocardial infarction
   - Nephrotic syndrome
   - Acute pancreatitis
   - Acute renal failure
5. **Different types of**
   - Acidosis
   - Alkalosis

Index